Digital Logic Design

W 830-1030
TA HRS F 7-11 ECA212

HW CHAP #

1 : 2, 3, 4, 5, 6, 9, 12, 13
2 : 6, 7, 9, 10, 13, 16   } DUE 9/9   55/60

3 : 3, 5, 8, 12
4 : 3, 4, 5, 8, 9, (1) 6   } DUE 9/16 TUES.   40/50

5 : 1, 2, 3, 4, 9
6 : 1, 2, 4, 5, 9   } DUE 9/18 TH.   15/35

7 : 2, 5, 6, 9
8 : 1, 4, 5, 9, 12   } DUE 9/25 TH 5pm   18/30

9 : 2, 5, 9, 10
10 : 1, 3, 4, 6   } DON'T TURN IN   73.%

11   10/10

12   10/10   96%

LAB #

  1 : HANDOUT   DUE WED 9/10   5 PM   ECA 235   (9.5)

  2 : TEXT CHAP 3 p 24 (LAB) DUE WED 9/17   (9.5)

  3 : NAND, NOR, MULTIPLEXERS   DUE   9/24

  4 : BINARY ADDERS, COMB. NTWK DUE   10/1

  5 : FLIP-FLOPS   10/8

COMPUTER ACCOUNTS - GET FROM LAB TA or MR. JAUREK

NEW TA (HW)   WED 1:30 to 3:00 pm } JUN MA

          FRI   9:00 to 11:00 AM } ECA 212

Digital Logic Design
Tutorials and
Laboratory Exercises

John F. Passafiume
Michael Douglas
Georgia Institute of Technology

1817

HARPER & ROW, PUBLISHERS, New York
Cambridge, Philadelphia, San Francisco,
London, Mexico City, São Paulo, Singapore, Sydney

Sponsoring Editor: John Willig
Project Editor: Nora Helfgott
Text Design Adaptation: Betty Sokol
Cover Design: Betty Sokol
Cover Photo: Stock, Boston
Text Art: Reproduction Drawings Limited
Production: Delia Tedoff
Compositor: ComCom Division of Haddon Craftsmen, Inc.
Printer and Binder: The Murray Printing Company

DIGITAL LOGIC DESIGN: Tutorials and Laboratory Exercises

Library of Congress Cataloging in Publication Data

Passafiume, John F.
  Digital logic design.

  Bibliography: p.
  1. Logic circuits. 2. Logic circuits—Laboratory
manuals. I. Douglas, Michael, 1960– . II. Title.
TK7868.L6P38 1985   621.3819′5835   84-10790
ISBN 0-06-045028-2

    86 87 9 8 7 6 5 4 3 2

Contents

Preface

This manual introduces the student to the fundamentals of digital logic design and circuitry. Early lab exercises deal with logic gates, truth tables, expression reduction, and implementation. Later labs employ these basic concepts to form important functional units such as multiplexors, decoders, adders, flip-flops, counters, and registers. This book is designed to stand alone, providing the textual support required to perform and understand the lab exercises. With this approach, the manual bridges the gap students often encounter when using separate text and lab books.

The material covered in *Digital Logic Design* is not, then, intended to replace a course that one would expect in an electrical engineering curriculum but rather to provide a working knowledge of the basic components of digital circuits, including, of course, digital computers. This level of exposure is ideal for students majoring in computer science or any technical area in which a familiarity with hardware can enhance one's performance. At Georgia Tech, the laboratory supports and supplements material presented in one of a sequence of three courses in computer organization and programming. The manual thus provides the opportunity to explain and demonstrate theoretical concepts in a laboratory environment.

We have chosen to use the TTL family of digital logic due to its availability and moderate cost. A suggested complement of TTL parts and laboratory equipment is provided in Appendix D. We have taken what we feel is a rather unique approach for a manual of this type. Rather than have students follow a cookbook list of steps to perform a laboratory exercise, we have arranged the exercises so that in most cases students are required to perform a simple design and then implement it with available hardware. We feel that this approach will enable students to gain an additional insight into the way the components work as well as an appreciation of the considerations to be addressed in digital logic design.

Each chapter is followed by a series of review questions designed to test students' understanding of the subject matter. Most of the questions can be answered by studying the text material in the chapter and extrapolating, when necessary, from principles and techniques covered therein. Some of the questions are in the "discovery" category and will normally require students to consult one or more of the references included in Appendix C. These questions will therefore extend and expand the coverage of digital logic design presented.

The lab course that this manual supports has been offered for two years at Georgia Tech and has been well received by the students there. Many of them have undertaken very ambitious projects as a result of this first exposure to the basic hardware elements that ultimately interpreted their programs. We would like to thank these students for their invaluable feedback, which contributed to revisions in our manuscript. Our work would not have been possible without the enthusiastic support of Dr. Ray Miller, our School Director, and Dr. Lucio Chiaraviglio, Associate Director and Head of the Computer Supported Instructional Group. Finally, we also wish to thank John Willig and the staff of Harper & Row for encouraging us to complete this work.

John F. Passafiume
Michael Douglas

Digital Logic Design

Review of Fundamental Concepts and An Introduction to the TTL Family

Objectives

1. A review of the basic logic gates and their functions.
2. An introduction to the TTL family—functional, electrical, and physical characteristics.
3. An introduction to circuit wiring and testing techniques.

1.1 Fundamental Digital Concepts

The student is expected to have had experience with boolean algebra and the basic logic gates. As a refresher we will review the five basic logic gates: AND, OR, INVERT, NAND, and NOR.

1.1.1 The AND Gate

The AND gate provides a true output only when all inputs are true. The nature of an AND gate is illustrated in Table 1.1. In this and the following truth tables, true is represented by a 1 and false is represented by a 0. This truth table shows the output F of a two-input AND gate for each possible combination of the inputs A and B. The American National Standards Institute (ANSI) symbol for the AND gate is shown in Figure 1.1.

In boolean algebra, the AND function is usually indicated by a dot between the

Table 1.1 Truth Table for the AND Function

A	B	F
0	0	0
0	1	0
1	0	0
1	1	1

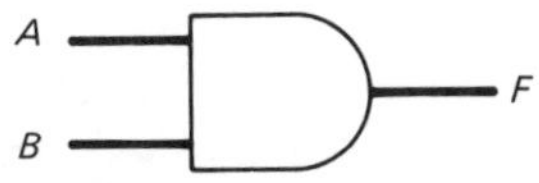

Figure 1.1 AND Gate

Table 1.2 Truth Table for the OR Function

A	B	F
0	0	0
0	1	1
1	0	1
1	1	1

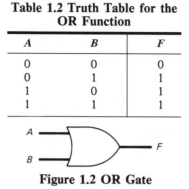

Figure 1.2 OR Gate

inputs (logical multiplication), or by implicit multiplication, as in algebra, by placing the input variables next to each other. For example:

$$F = A \text{ AND } B$$

can be written as:

$$F = A \cdot B$$
$$F = AB$$
$$F = (A)(B)$$

1.1.2 The OR Gate

The OR gate provides a true output when any of its inputs are true. The nature of an OR gate is illustrated in Table 1.2. This truth table shows the output F of a two-input OR gate for each possible combination of the inputs A and B. The ANSI symbol for the OR gate is shown in Figure 1.2.

In boolean algebra, the OR function is usually indicated by a plus sign (logical addition) between the inputs. For example:

$$F = A \text{ OR } B$$

can be written as:

$$F = A + B$$

1.1.3 The NOT Gate or Inverter

The output of an inverter is simply the complement of the input. The nature of an inverter is illustrated in Table 1.3, which shows the output of an inverter for the two possible inputs. The ANSI symbol for the inverter is shown in Figure 1.3.

Table 1.3 Truth Table for the NOT Function

A	F
0	1
1	0

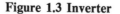

Figure 1.3 Inverter

Table 1.4 Truth Table for the NAND Function

A	B	F
0	0	1
0	1	1
1	0	1
1	1	0

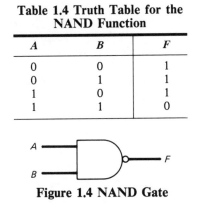

Figure 1.4 NAND Gate

In boolean algebra, the NOT function is usually indicated by a bar over the variable or a following prime. For example:

$$F = \text{NOT } A$$

can be written as:

$$F = \overline{A}$$
$$F = A'$$

The use of the prime should be limited to typewriters where the bar cannot be generated. The bar is much more readable, especially in more complex expressions.

1.1.4 The NAND Gate

The output of the NAND gate is simply the complement of the output of the AND gate. Thus the NAND gate provides a false output only when all inputs are true. The nature of the NAND gate is illustrated in Table 1.4. This truth table shows the output F of a two-input NAND gate for all possible combinations of the inputs A and B. The ANSI symbol for the NAND gate is shown in Figure 1.4.

In boolean algebra, the NAND function is not a primitive operator but a combination of the AND and invert functions. A NAND operation is simply the AND function with its output inverted: (N)ot (AND). This can be illustrated in boolean algebra as:

$$F = \overline{AB}$$
$$F = \overline{A \cdot B}$$

1.1.5 The NOR Gate

The output of the NOR is the complement of the output of the OR gate. Therefore, the output of the NOR gate is illustrated in Table 1.5. This truth table shows the output F of a two-input NOR gate for all possible combinations of the inputs A and B. The ANSI symbol for the NOR gate is shown in Figure 1.5.

In boolean algebra, the NOR function is not a primitive operator but a combina-

Table 1.5 Truth Table for the NOR Function

A	B	F
0	0	1
0	1	0
1	0	0
1	1	0

Figure 1.5 NOR Gate

tion of the OR and invert functions. A NOR operation is simply the OR function with its output inverted: (N)ot (OR). This can be illustrated in boolean algebra as:

$$F = \overline{A + B}$$

1.2 Introduction to the TTL Family

So far we have talked of the five basic gates in terms of true and false, logic symbols, and truth tables. But how are these logic gates used in actual circuits? Logic gates can be built in a variety of ways. Some of the first computing devices used mechanical relays to implement logic gates. More recent forms include metal-oxide semiconductor (CMOS, PMOS, NMOS), emitter-coupled logic (ECL), resistor-transistor logic (RTL), diode-transistor logic (DTL), and transistor-transistor logic (TTL). TTL is the most popular logic family today and will be used in this course.

1.2.1 TTL Characteristics

The members of the TTL family are each identified by a unique part number. TTL part numbers are in the form 74XXX, where the XXX is a two- or three-digit number uniquely identifying the particular type of chip. For example, the part number for the TTL chip containing four two-input AND gates is 7408. The chip containing four two-input OR gates is the 7432.

In addition to the 74 series, there is a 54 series of chips that is functionally equivalent to the corresponding 74 series of chips but meet more stringent military specifications.

In the TTL family, a high logic level is represented by a voltage in the range of 2.0v–5.0v, and a low logic level by a voltage in the range of 0.0v–0.8v. Any voltage greater than 0.8v and less than 2.0v is undefined. Typically, a high logic level represents a 1 or TRUE and a low logic level represents a 0 or FALSE. These levels are referred to as *active high* logic levels. With *active low* logic levels, a low logic level represents a 1 or TRUE and a high logic level represents a 0 or FALSE.

1.2.2 TTL Subfamilies

Within the TTL family, several subgroupings are present. For example, chips are available with one of three different output circuits: totem-pole, open-collector, and three-state.

Totem-pole outputs derive their name from the appearance of the output circuit. A pull-up transistor and pull-down transistor are "stacked" on top of each other to decrease switching time and reduce the number of external components required to connect gates. Never connect the outputs of totem-pole gates together; the connection may damage the gates. We will deal primarily with totem-pole outputs in this course.

Open-collector outputs have no pull-up transistor. Instead, an external pull-up resistor is required for the gate to assume a high logic level. This makes open-collector gates ideal for interface applications since the pull-up voltage can usually be up to 15 or 30 volts. Also, the outputs of open-collector gates can be connected together to form hard-wired logic.

Three-state outputs can assume one of three states: a high logic level, a low logic level, and a third state that appears as an open circuit—as if nothing is there. This third state is often used in computers where many devices must access a single data bus.

There are several different internal constructions (affecting the speed and power consumption of the chip) that serve to further classify the TTL family. In addition to the standard 74 series, there is a low-power, low-speed series, 74L; a high-power, high-speed series, 74H; another high-power yet even faster series using Schottky diodes, 74S; and, finally, a very low-power, yet faster-than-standard TTL series using Schottky diodes, the 74LS series.

1.3 Wiring and Testing a Circuit

In later labs, more and more complex circuits will be designed and wired. Following the basic guidelines provided here will make the wiring and debugging effort much simpler.

The first requirement in implementing a circuit after it has been designed is obtaining the necessary parts. The proper TTL chips to perform the logic functions required can be found by looking in Appendix B or in a TTL data book. For example, if a two-input NAND gate is needed, we find the proper TTL part number is 7400.

Once the part numbers are obtained, it is wise to write the pin numbers for each input and output used on the chip on your logic diagram. This will make wiring much simpler and reduce the chance for error. In addition to gate connections, each TTL chip will require two additional connections: a +5v power-supply connection and a ground connection. Typically, pin 14 is connected to +5v and pin 7 is connected to ground. Be careful though; there are exceptions.

It is very important to properly identify pin numbers before wiring a circuit. A chip can be ruined by simply inserting it in a circuit backwards. Figure 1.6 shows the pin numbering of a typical TTL chip. With the chip oriented as shown, pin 1 will always be the top left pin.

1.3.1 Wiring the Circuit

Circuits will be wired on a logic design station. These design stations generally provide a breadboard for wiring the circuit, a power supply, switches for inputs, LEDs for output monitoring, and, finally, some sort of debounced switch inputs. The

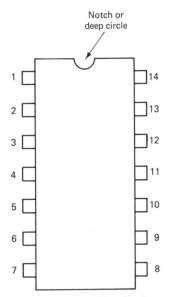

Figure 1.6 Pin Numbering

use and features of the logic designer you will use will be covered during the first lab session.

Wiring a circuit, especially a complex one, can be very frustrating. To make the process easier and reduce the chance for error, here are a few simple guidelines that should be followed.

1. Always start with the power OFF on the logic designer.
2. Insert the required chips in the breadboard. (Make sure the chips are oriented correctly.)
3. Connect all power (Vcc) and ground pins to +5v and ground respectively.
4. Connect wires between pins, switches, LEDs, and so on, as labeled on your logic diagram.
5. Visually inspect your circuit and correct any mistakes.
6. Turn on the power and test the circuit. The use of test equipment such as the logic probe, oscilloscope, and volt meter will be covered in the lab.

Lab Exercise

Objective

This lab is intended to introduce the student to the use of the lab equipment: the logic designer, the test equipment, and, of course, the TTL chips.

Procedure

This lab begins with a brief lecture introducing the procedures of the lab and the use of the lab equipment.

1. AND gate operation.
   (a) Look up the TTL part number for a two-input AND gate, and obtain this chip from your lab instructor.
   (b) Connect the inputs of one of the AND gates to the switches on the logic designer and the output of that gate to an LED on the logic designer.
   (c) Verify proper operation of the AND gate for all input combinations by using the input switches. When the LED is lit, this indicates a high logic level or 1. When the LED is off, this indicates a low logic level or 0.
   (d) Record the TTL part number, and produce a truth table based on the results of (c).
2. OR gate operation. (Repeat (1) for a two-input OR gate.)
3. NAND gate operation. (Repeat (1) for a two-input NAND gate.)
4. NOR gate operation. (Repeat (1) for a two-input NOR gate.)
5. What level does an unconnected input appear to be (high or low)? (*Hint:* Repeat one of the previous operations with one input left unconnected.) What reasoning did you use to determine your answer?

Review Questions

1.1 Generate the truth table for:
   (a) A four-input NAND gate
   (b) A four-input NOR gate
1.2 Can a NAND gate be used as an INVERTER? If so, illustrate how with a two-input NAND gate.
1.3 Can a NOR gate be used as an INVERTER? If so, illustrate how with a two-input NOR gate.
1.4 Assuming POSITIVE logic, what do each of the following voltage levels represent (1 or 0)?
   (a) 0.3v
   (b) 2.4v
   (c) 5.0v

(d) 0.0v

(e) 1.9v

1.5 Find the TTL part number and the number of gates per chip for each of the following logic gates:

(a) Inverters

(b) Two-input OR gates

(c) Four-input NAND gates

(d) Two-input NOR gates

(e) Two-input AND gates

1.6 What pin numbers must be connected to power supply and ground for each of the following TTL chips:

(a) 7400

(b) 7420

(c) 7476

(d) 74151

1.7 Which of the following TTL subfamilies is fastest in switching speed?

(a) 74XX

(b) 74SXX

(c) 74LSXX

1.8 Which of the following TTL subfamilies is lowest in power consumption?

(a) 74XX

(b) 74LSXX

(c) 74SXX

1.9 Show that the distributive law of OR over AND holds true by means of a truth table; that is, show that:

$$X + (Y \cdot Z) = (X + Y) \cdot (X + Z)$$

Note that this is not valid for ordinary algebra.

1.10 Show that the distributive law of AND over OR holds true by means of a truth table; that is, show that:

$$X \cdot (Y + Z) = (X \cdot Y) + (X \cdot Z)$$

Note that this rule holds for ordinary algebra.

1.11 Use a truth table to prove the absorption rule; that is,

$$X + X \cdot Y = X$$

1.12 Use a truth table to prove the following theorem.

$$X \cdot (X + Y) = X$$

1.13 Given the following TTL gates and connections, give an expression for the output function F.

(a)

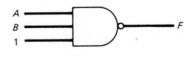

(b)

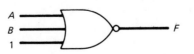

1.14 Which of the following families has the fastest switching speed?
   (a) ECL
   (b) CMOS
   (c) 74SXX
   (d) DTL

1.15 Which of the following families has the lowest standby current drain?
   (a) ECL
   (b) CMOS
   (c) TTL
   (d) DTL

1.16 Which of the following TTL subfamilies is typically lowest in cost?
   (a) 74XX
   (b) 74SXX
   (c) 74LSXX

1.17 What would be the result of connecting two inverters in tandem? Why would this be a useful thing to do in a digital circuit?

1.18 What is the disadvantage of leaving the inputs of a logic gate unconnected?

chapter 2

Basic Two-Level Circuits

Objectives

1. An introduction to standard two-level forms: sum of products (SOP) and product of sums (POS).
2. An introduction to methods for implementing simple SOP and POS equations.

2.1 Sum of Products Form

Assume the existence of a simple digital network and that the operation of this network is defined by Table 2.1. This network has three inputs: A, B, and C, and an output F. The operation of this network can readily be determined by observing the truth table. The output F can be seen to equal 1 in four cases:

1. When $A = 0$ and $B = 1$ and $C = 0$ or
2. When $A = 0$ and $B = 1$ and $C = 1$ or
3. When $A = 1$ and $B = 0$ and $C = 0$ or
4. When $A = 1$ and $B = 1$ and $C = 1$

Assuming the true state of a variable to be 1 and the complemented state of a variable to be 0, the preceding observation can be converted almost directly into the following boolean expression:

$$F = \overline{A}B\overline{C} + \overline{A}BC + A\overline{B}\overline{C} + ABC \qquad (2.1)$$

This form of an equation, a series of product (AND) terms connected by addition (OR), is referred to as a *sum of products* (SOP) equation. This form is the most easily derived from a truth table.

Table 2.1

A	B	C	F
0	0	0	0
0	0	1	0
0	1	0	1
0	1	1	1
1	0	0	1
1	0	1	0
1	1	0	0
1	1	1	1

2.1.1 SOP Circuit Implementation

Once the proper SOP expression has been derived, the next step is the design of a logic diagram from which a circuit can be built to implement the digital network.

An expression in SOP form is easily converted into a logic diagram. The result of an SOP equation is the OR of the results of several AND functions. Thus the output of the logic network should also be the result of an OR operation on the results of several AND functions. The logic network to implement Equation 2.1 is illustrated in Figure 2.1.

As shown here, the output F is the result of an OR operation on the results of several AND functions—a direct parallel to the original SOP equation. Also note the two-level nature of the logic diagram. Ignoring the input inverters, we see that the longest path is always just two gates: the AND then the OR. Circuits in this form are referred to as two-level circuits.

In this diagram, the inverters providing the complements of the input variables are shown explicitly. Often these input inverters are not shown and the inputs to each gate simply labeled with the proper value. This simplifies the logic diagram and makes the inputs to the gates more readily apparent. Figure 2.2 (page 13) shows this format. When it is assumed that both a variable and its complement are available, as in Figure 2.2, we refer to these as *double-rail* inputs. Assume double-rail inputs are available on all logic diagrams you produce unless specifically stated otherwise.

2.1.2 Normal Form

An SOP equation in which each product (AND) term contains all input variables in either true or complemented form is said to be in *normal* or *canonical* form. Thus Equation 2.1 is in normal form since each product term contains all three input variables (A, B, and C) in either true or complemented form.

An equation in normal form will have one product term for each entry in the truth table for which the output is 1. For example, Equation 2.1 contains four product terms,

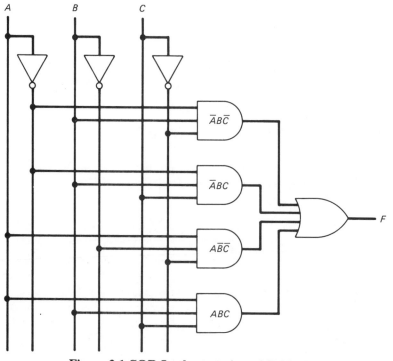

Figure 2.1 SOP Implementation of Table 2.1

and in the truth table there are four entries for which the output equals 1. If there were just three entries for which the output equaled 1, then the normal form equation for that truth table would contain just three product terms.

An equation in normal form is not necessarily the simplest or only equation to evaluate a particular boolean function. For example, the following boolean equation also evaluates the function defined by Table 2.1:

$$F = \overline{A}B + A\overline{B}\,\overline{C} + BC$$

This equation is not in normal form since not all product terms contain all input variables. The first term is missing input C, and the last term is missing input A.

Though there are typically several equations that will evaluate a given function, there is just one equation in normal form that will. Thus, to remove ambiguity, a function is often expressed in normal form. To convert an SOP equation to normal form, terms with missing input variables must be "multiplied" by the missing variable. Since multiplication (ANDing) by 1 will not change a product term, we could rewrite the previous equation as:

$$F = \overline{A}B(1) + A\overline{B}\,\overline{C} + (1)BC$$

From the laws of boolean algebra we know that ORing any variable with its complement always equals 1:

$$(X + \overline{X}) = 1$$

Thus we can rewrite the previous equation as:

$$F = \overline{A}B\,(C + \overline{C}) + A\overline{B}\,\overline{C} + (A + \overline{A})BC$$

This provides the missing input variables for the product terms. Simplifying the previous equation yields:

$$F = \overline{A}BC + \overline{A}B\overline{C} + A\overline{B}\,\overline{C} + ABC + \overline{A}BC$$

Since the first and last terms are the same, one can be removed leaving the normal form equation

$$F = \overline{A}B\overline{C} + \overline{A}BC + A\overline{B}\,\overline{C} + ABC$$

which is the same as the original Equation 2.1.

2.1.3 Minterms

Looking at Table 2.1 we see that each combination of input variables forms a binary value between 0 and $2^x - 1$ where N equals the number of input variables. In the previous example, N is 3; thus possible values range from 0 through 7 (000–111). Since each possible combination of input variables forms a unique binary value, a particular input combination can be represented by the decimal equivalent of the binary value formed by the inputs. This value is referred to as the *minterm* number.

For example, what input combination does minterm 5 ($m5$) represent? Assuming the three input variables used previously:

$$5 \text{ decimal} = 101 \text{ binary} => A\overline{B}C$$

What input combination is represented by minterm 2?

$$2 \text{ decimal} = 010 \text{ binary} => \overline{A}B\overline{C}$$

What minterm represents the input combination $AB\overline{C}$?

$$AB\overline{C} => 110 \text{ binary} = 6 \text{ decimal}$$

What minterm represents the input combination $A\overline{B}\,\overline{C}$?

$$A\overline{B}\,\overline{C} => 100 \text{ binary} = 4 \text{ decimal}$$

Using minterms to represent input combinations, Equation 2.1 can be rewritten as:

$$F = m2 + m3 + m4 + m7$$

This equation can be further simplified into "minterm shorthand" form by using summation:

$$F = \Sigma m(2, 3, 4, 7)$$

where the small m indicates that the numbers represent the minterm numbers of the input combinations to be ORed.

2.2 Product of Sums Form

Where the SOP equation is a series of products (ANDs) connected by addition (OR), the product of sums equation (POS) is a series of sums (ORs) connected by multiplication (AND). Though derivation of a POS equation from a truth table is straightforward, the process is not as clear as with the SOP form.

The output of a digital network will equal 1 only when it is not equal to 0. This is rather obvious, but it is the basis of a POS equation. In Table 2.1, the output equals 0 for four input combinations: minterms 0, 1, 5, and 6. Thus we can say the output will equal 1 if and only if the input combination is not minterm 0 and not minterm 1 and not minterm 5 and not minterm 6; thus:

$$F = \overline{m0} \cdot \overline{m1} \cdot \overline{m5} \cdot \overline{m6} \qquad (2.2)$$

But what does "not minterm n" mean? The answer to this is illustrated in the following sequence:

$$m5 => A\overline{B}C$$
$$\overline{m5} => \overline{(A\overline{B}C)}$$

Apply DeMorgan's theorem:

$$\overline{m5} => (\overline{A} + \overline{\overline{B}} + \overline{C})$$
$$\overline{m5} => (\overline{A} + B + \overline{C})$$

This of course forms the sum terms of a POS equation. We can now write a complete POS equation for Table 2.1 as:

$$F = (A + B + C) \cdot (A + B + \overline{C}) \cdot (\overline{A} + B + \overline{C}) \cdot (\overline{A} + \overline{B} + C) \quad (2.3)$$

The procedure for deriving a POS equation can be summarized as follows: The POS equation will be the logical product of N sum terms where N is the number of truth-table entries for which the output is 0. The sum terms are formed by ORing the complements of each input variable for those input combinations where the output of the function is 0.

Since Table 2.1 has four entries for which the output is 0, there are four sum terms in the POS equation. If there were five entries in the truth table for which the output was 0, the POS equation would have five sum terms.

2.2.1 POS Circuit Implementation

Having derived a POS equation from the truth table, we find that the design of a logic diagram to implement that equation is similar to the process used for an SOP equation. The result of a POS equation is the ANDing of the results of several OR functions. Thus the output of the logic network designed should also be the result of an AND operation on the results of several OR functions. The logic diagram to implement Equation 2.3 is illustrated in Figure 2.2.

As shown here, the output F is the result of an AND operation on the results of several OR functions—a direct parallel to the original POS equation. Again note the two-level nature also characteristic of the SOP implementation and the use of double-rail inputs.

2.2.2 Maxterms

When the POS equation was derived, the inverted minterm was used for the sum terms. An inverted minterm is called a *maxterm.* Equation 2.2 was originally written as:

$$F = \overline{m0} \cdot \overline{m1} \cdot \overline{m5} \cdot \overline{m6}$$

Using "maxterm shorthand," this equation can be further simplified as:

$$F = \prod \overline{m}(0, 1, 5, 6)$$

where the $\overline{m}$ indicates the numbers represent maxterms.

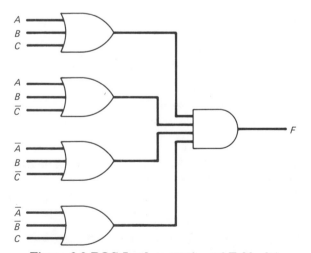

Figure 2.2 POS Implementation of Table 2.1

2.3 Design Considerations

Though SOP implementations are used most often, there are applications better suited to a POS implementation. For example, consider a three-input digital network that has an output of 1 for six of the eight possible input combinations. Expressed in SOP form, the resulting equation would have six terms and result in a circuit requiring six AND gates and a six-input OR gate. A POS implementation of the same network would have only two terms resulting in a circuit requiring just two OR gates and a two-input AND gate.

At this point it seems SOP or POS should be chosen based on whether there are more 1s or 0s in the output of the function (more 1s for POS, more 0s for SOP). Later labs will reveal that the decision is not that simple since expressions can typically be reduced to simpler form. In addition, SOP and POS are each individually suited to implementation with just NAND gates (SOP) or just NOR gates (POS). In this case, the implementation chosen might depend on gate availability, price, and so on.

Lab Exercise

Objectives

The first part of this lab lets the student implement a simple SOP and POS equation. The second part of this lab extends slightly beyond material presented requiring the implementation of a more complex SOP circuit using only two-input gates.

Procedure

1. The truth table below illustrates the Exclusive OR function. Produce the following SOP implementations for this function:
   (a) SOP equation
   (b) Minterm shorthand equation
   (c) Logic diagram
   (d) TTL part numbers to implement (c)

A	B	F
0	0	0
0	1	1
1	0	1
1	1	0

2. Produce the following POS implementations for the function shown in the table in (1).
   (a) POS equation
   (b) Maxterm shorthand equation
   (c) Logic diagram
   (d) TTL part numbers to implement (c)
3. Wire each of your designs from (1) and (2). Inputs A and B can be obtained from the switches on your logic designer. Use inverters to generate the complements of the input variables as required. Connect the output of the circuit to an LED on your logic designer and verify operation of your circuit with the truth table in (1).
4. Design an SOP implementation of the network defined in Table 2.1 using only inverters, two-input AND gates, and two-input OR gates:
   (a) Produce a logic diagram similar to Figure 2.1 using only inverters, two-input AND gates, and two-input OR gates.
   (b) Wire your design and verify its operation.

Have the lab instructor check your work after each circuit is working.

Review Questions

2.1 Convert the following expressions to normal form:
(a) $F(A, B) = B$
(b) $F(A, B, C) = \overline{A}B + B\overline{C}$
(c) $F(A, B, C) = A + B\overline{C}$
(d) $F(A, B, C) = A + C$
(e) $F(A, B, C) = ABC + A + BC$
(f) $F(A, B, C) = AB + BC$
(g) $F(A, B, C, D) = AB + AC$
(h) $F(A, B, C, D) = \overline{B}\overline{D} + \overline{A}C$

2.2 Expand the following expression into normal form and express as a sum of products.

$$F(X, Y, Z) = XYZ + \overline{X}\overline{Y}Z + Y\overline{Z}$$

2.3 Represent the expression in Question 2.2 in product-of-sums form.
2.4 Implement the expression in Question 2.2 using AND and OR gates.
2.5 Implement the resulting POS equation from Question 2.3 using AND and OR gates.
2.6 Express the following equation in SOP form:

$$F(A, B, C) = (A + B + C) \cdot (A + \overline{B} + C) \cdot (\overline{A} + B + \overline{C})$$

2.7 Express the following in minterm shorthand form:

$$F(A, B, C, D) = ABD + \overline{A}BC + A\overline{B}D + \overline{B}CD$$

2.8 Express the following in maxterm shorthand form:

$$F(A, B, C, D) = (\overline{A} + \overline{B} + \overline{D}) \cdot (A + B + \overline{C}) \cdot (\overline{A} + B + D) \cdot (B + \overline{C} + \overline{D})$$

2.9 Produce for the following truth table:
(a) SOP equation for F
(b) Minterm shorthand equation for F
(c) Logic diagram for SOP implementation

A	B	C	F
0	0	0	0
0	0	1	1
0	1	0	1
0	1	1	0
1	0	0	1
1	0	1	0
1	1	0	0
1	1	1	1

2.10 Repeat Question 2.9 producing a POS implementation to include:
(a) POS equation for F
(b) Maxterm shorthand equation for F
(c) Logic diagram for POS implementation
2.11 Implementing the function in Question 2.2 might be difficult since there are no standard TTL OR gates available with more than two inputs. Produce logic diagrams illustrating how to make the following:
(a) A three-input OR gate using two-input OR gates
(b) A four-input OR gate using two-input OR gates

2.12 There are AND gates available with more than two inputs, but if only two-input AND gates were available, illustrate how to make the following:

    (a) A three-input AND gate using two-input AND gates

    (b) A four-input AND gate using two-input AND gates

2.13 Convert the following to the other normal form:

    (a) $F(x, y, z) = \sum m(1, 3, 7)$

    (b) $F(A, B, C, D) = \sum m(0, 2, 6, 11, 13, 14)$

    (c) $F(x, y, z) = \prod \overline{m}(0, 3, 6, 7)$

    (d) $F(A, B, C, D) = \prod \overline{m}(0, 1, 2, 3, 4, 6, 12)$

2.16 Express the following functions in SOP and POS forms:

    (a) $F(A, B, C, D) = D(\overline{A} + B) + \overline{B}D$

    (b) $F(w, x, y, z) = \overline{y}z + wx\overline{y} + wx\overline{z} + \overline{w}\,\overline{x}z$

    (c) $F(A, B, C) = (\overline{A} + B) \cdot (\overline{B} + C)$

    (d) $F(x, y, z) = (xy + z)(y + xz)$

2.17 Given the following boolean functions, F and G, obtain the truth table for $F + G$.

$$F = D + AB\overline{C} + \overline{A}C$$
$$G = D(\overline{A} + \overline{B} + C)(A + \overline{C})$$

3

Implementation with One Gate Type

Objective

An introduction to two methods for implementing a digital circuit using only one gate type: algebraic manipulation and gate equivalency rules.

3.1 Motivation

In industry, cost is typically the controlling factor in the design and production of goods. In producing digital circuits, even though it may seem that the purchasing department is responsible for reducing costs by finding the cheapest supplier, engineers designing digital circuits have several methods at their disposal for reducing the cost of the design.

In general, the fewer and less expensive the parts required to build a circuit, the less it will cost. Therefore, an important goal in circuit design is minimizing the hardware required and the associated cost of that hardware.

In Chapter 2's lab, simple guidelines were introduced for converting an SOP or POS expression into a logic diagram. For example, consider the following SOP expression:

$$F = A\overline{B} + \overline{C}D \tag{3.1}$$

Implementing this circuit would require two AND gates and one OR gate. The TTL chips required are the 7408 (four ANDs) and the 7432 (four ORs). Implementing the preceding expression would use just two of the four AND gates and just one of the four OR gates. In production, this kind of inefficiency would be very costly.

To get around this inefficiency, it is common practice to implement an expression using only one type of gate—typically a NAND or NOR gate. The NAND gate is the most "naturally" implemented function at the internal level of the chip. This quality results in the NAND gate being the fastest and least expensive of the TTL chips. In addition, both the NAND and the NOR gate can act as inverters, thus eliminating the need for separate inverter chips.

How can an expression in terms of ANDs and ORs be realized using only NAND or NOR gates? There are two primary methods for accomplishing this. First, using boolean algebra, the original expression can be manipulated into an equivalent expression using the gate type desired. This method can become extremely complicated, especially when dealing with complex expressions. A second method uses gate equivalency rules that let any gate perform the AND and OR functions.

3.2 Algebraic Manipulation

The laws of boolean algebra allow us to convert a boolean expression into an equivalent boolean expression that is directly implementable in the gate type desired. For example, consider implementing Equation 3.1 using NAND gates only. The function of the NAND gate on two inputs, A and B, is illustrated in boolean algebra as:

$$F = \overline{AB}$$

Thus in implementing an expression using only NAND gates, the goal is to reduce all logical operations to the form above—that of a NAND gate. From our original expression:

$$F = A\overline{B} + \overline{C}D$$

Double bar and apply DeMorgan's theorem to make all operations in the AND form:

$$F = \overline{\overline{A\overline{B} + \overline{C}D}}$$
$$F = \overline{(\overline{A\overline{B}}) \cdot (\overline{\overline{C}D})}$$

Examining this expression we find three NAND forms:

1. $\overline{(A\overline{B})}$
2. $\overline{(\overline{C}D)}$
3. $\overline{(XY)}$ where $X = \overline{(A\overline{B})}$ and $Y = \overline{(\overline{C}D)}$

Thus the expression can be realized using three NAND gates as illustrated in Figure 3.1.

What if there were a surplus of NOR gates on hand; how could Equation 3.1 be implemented using NOR gates? Following the same procedure used with the NAND gate, we recall the function of the NOR gate on two inputs, A and B, is illustrated in boolean algebra as:

$$F = \overline{A + B}$$

Therefore, to implement an expression using only NOR gates, all logical operations must be reduced to the preceding form—that of a NOR gate. From the original expression:

$$F = A\overline{B} + \overline{C}D$$

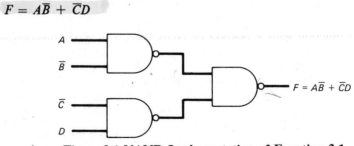

Figure 3.1 NAND Implementation of Equation 3.1

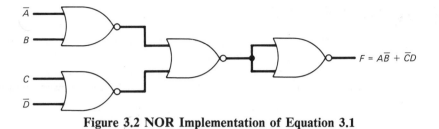

Figure 3.2 NOR Implementation of Equation 3.1

Double bar and apply DeMorgan's theorem to make all logical operations in the OR form.

$$F = \overline{\overline{A\overline{B} + \overline{C}D}}$$
$$F = \overline{\overline{(A\overline{B})\cdot(\overline{C}D)}}$$
$$F = \overline{\overline{(A\overline{B})} + \overline{(\overline{C}D)}}$$
$$F = \overline{\overline{(\overline{A} + B)} + \overline{(C + \overline{D})}}$$

Examining this expression we find two NOR forms:

1. $\overline{(\overline{A} + B)}$
2. $\overline{(C + \overline{D})}$

The final logical operation between (1) and (2) is the OR function. If a NOR gate is used to perform the OR operation, the final output will be inverted. Therefore an inverter must be placed at the output of this final NOR gate to properly generate *F*. Note the use of the NOR gate as an inverter in the logic diagram in Figure 3.2. Again the goal is using each gate on each chip.

The possible complexity of using algebraic manipulation is obvious. With this complexity comes a high probability for error. Because of these problems, we study another method for realizing an expression in just one gate type using gate-equivalency rules.

3.3 Gate Equivalencies

Boolean algebra is composed of just three operators: AND, OR, and NOT. Any expression can be implemented using combinations of these three functions. To implement a circuit with just one gate type, all three of these operations must somehow be performed.

3.3.1 Performing the NOT Function with NAND and NOR

Chapter 1 presented the problem of performing the NOT function with NAND and NOR gates as an exercise. NAND and NOR can function as inverters by tying the inputs together as shown in Figure 3.3.

3.3.2 Performing the AND Function with NAND and NOR

Through simple boolean transformations we can determine how to perform the AND function with a NAND or NOR gate. The AND function on two inputs is illustrated in boolean algebra as:

$$F = AB$$

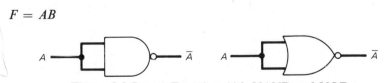

Figure 3.3 Invert Function with NAND and NOR

The NAND function is expressed as:

$$F = \overline{AB}$$

This equation can be reduced to the AND function by complementing the entire equation (i.e., the output) to remove the NOT bar. Thus a NAND gate will perform the AND function if the output is inverted. The NOR function on two inputs is illustrated as:

$$F = \overline{A + B}$$

By applying DeMorgan's theorem, we obtain an equivalent expression:

$$F = \overline{A}\,\overline{B}$$

This can be reduced to the AND function by complementing the input variables. Thus a NOR gate will perform the AND function if the inputs are inverted.

In summary, a NAND gate will perform the AND function if the output is inverted. A NOR gate will perform the AND function if the inputs are inverted.

3.3.3 Performing the OR Function with NAND and NOR

The OR function on two inputs is illustrated in boolean algebra as:

$$F = A + B$$

The NAND function is expressed as:

$$F = \overline{AB}$$

By applying DeMorgan's theorem we obtain an equivalent expression:

$$F = \overline{A} + \overline{B}$$

This can be reduced to the OR function by complementing the input variables. Thus a NAND gate performs the OR function if the input variables are complemented. The NOR function on two inputs is illustrated as:

$$F = \overline{A + B}$$

This can be reduced to the OR function by complementing the entire equation (i.e., the output) to remove the NOT bar. Thus a NOR gate performs the OR function if the output is complemented.

In summary, a NAND gate will perform the OR function if the inputs are inverted. A NOR gate will perform the OR function if the output is inverted.

3.3.4 Symbolic Definitions

To clarify this dual nature of gates, each can be represented symbolically in two forms. Shown in Figure 3.4 and Figure 3.5 are two representations of a NAND gate and two representations of a NOR gate respectively. The intended function (AND or OR) is represented by the basic shape of the symbol. Input or output requirements are dictated by the presence of circles at the inputs or outputs. In direct correlation with the summaries presented in Sections 3.3.2 and 3.3.3 we note that the presence of a circle at an input indicates that the input must be inverted to properly perform the function

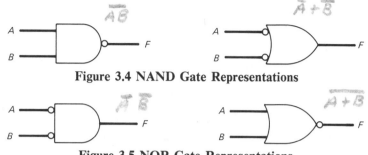

Figure 3.4 NAND Gate Representations

Figure 3.5 NOR Gate Representations

dictated by the symbol shape, that is, the AND or OR function. Also, the presence of a circle at an output indicates that the output must be inverted to properly perform the function dictated by the symbol shape, that is, the AND or OR function.

It is important to note that these symbols do not represent new gates, just two ways of viewing the NAND and NOR gates. Thus we can speak of the *AND form* of a NAND gate and the *OR form* of a NAND gate. Likewise we can speak of the *AND form* of a NOR gate and the *OR form* of a NOR gate.

3.3.5 Active High and Active Low Inputs and Outputs

To further define the meaning of the circles on the gate symbols, we say a gate form with a circle at the output has an *active low output*. A gate form with circles at the inputs is said to have *active low inputs*. Conversely, a gate form without a circle at the output is said to have an *active high output*, and a gate form without circles at the inputs is said to have *active high inputs*.

For example, the AND form of the NAND gate can be thought of as an AND gate with active high inputs and an active low output. The OR form of the NAND gate can be thought of as an OR gate with active low inputs and an active high output.

The term *active high* refers to logic levels as we have dealt with them so far: a true is represented by a high logic level. *Active low* indicates that a true is represented by a low logic level rather than a high logic level. To illustrate, consider performing the AND function with a NAND gate. The AND function provides a true output only if all inputs are true. Looking at the AND form of the NAND gate, we see it too performs the AND function, but since it has an active low output, a true output is indicated by a low logic level rather than a high logic level. Thus to see a *normal* active high output, we must invert the output of the NAND gate—just as was stated in Section 3.3.2.

An OR gate produces a true output if any of its inputs are true. Looking at the OR form of the NAND gate, we see it also performs the OR function, but since it has active low inputs, a true input is indicated by a low logic level rather than a high logic level. To provide these active low inputs from *normal* active high inputs, we must invert each of the inputs. Note this same requirement stated in Section 3.3.3.

Take your time studying the truth tables for these gates in Chapter 1, and convince yourself the relations stated previously hold.

3.4 Logic Diagram Design Using Gate Equivalencies

Keeping in mind the preceding introduction to the concept of gate equivalencies, we now turn to the design of a logic diagram using this concept. Using the guidelines from Chapter 2, the logic diagram to implement Equation 3.1 looks like Figure 3.6.

3.4.1 Logic Diagram Design Using NAND Gates

This diagram is easily converted to use just NAND gates by first replacing all OR gates with the OR form of the NAND gate and replacing all AND gates with the AND form of the NAND gate, as in Figure 3.7. Do not draw interconnections yet.

$F = A\overline{B} + \overline{C}D$

Figure 3.6 AND-OR Implementation of Equation 3.1

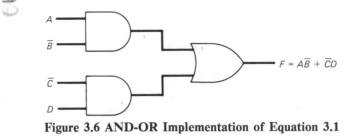

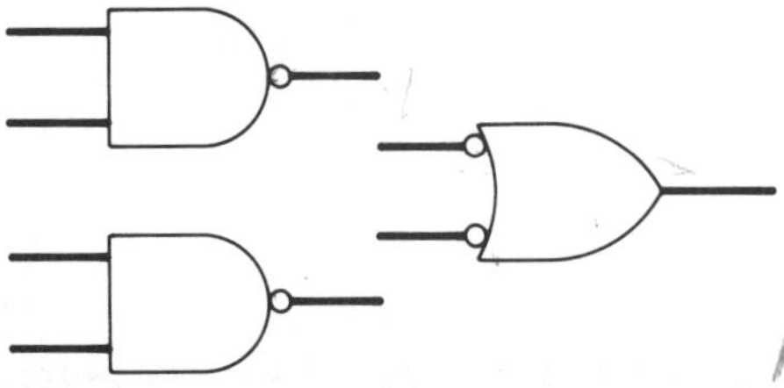

Figure 3.7 Replacing with AND and OR Form of AND Gate

Since we deal with active high inputs and outputs in the "real" world, the initial inputs and final outputs of a circuit should generally be active high. Looking at Figure 3.7 we see that this condition is already satisfied. The next step is connection of the AND forms to the OR form. Since the OR form of the NAND gate requires active low inputs and the output of the AND form is already active low, no level conversion is needed, and a direct connection can be made between the outputs of the AND forms and the inputs of the OR form. In general, a direct connection can be made between an active low output and input or between an active high output and input. If one is active high and the other is active low, then an inverter must be inserted to convert from one level to the other. Making connections as just described, the final logic diagram using only NAND gates is illustrated in Figure 3.8. When the AND and OR forms of the NAND gates on the logic diagram are shown, the intended function of the circuit is more obvious. Looking at Figure 3.8, we can readily see it is a SOP implementation—a series of ANDs feeding an OR—while looking at the equivalent circuit in Figure 3.1, we find that the actual function of the circuit is not clear.

A second example will illustrate a condition where level conversion is necessary. Consider implementing a three-input OR gate using only two-input NAND gates. If

$F = A\overline{B} + \overline{C}D$

SAME AS FIG 3.1

Figure 3.8 AND Implementation of Equation 3.1

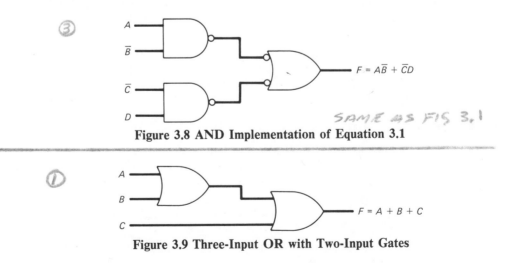

$F = A + B + C$

Figure 3.9 Three-Input OR with Two-Input Gates

two-input OR gates were available, this function could be implemented as shown in Figure 3.9.

To implement using only two-input NAND gates we proceed as before replacing all OR gates with the OR form of the NAND gate. This yields Figure 3.10. To complete the diagram we must first insure active high inputs and outputs to the real world. Since the inputs of the OR forms are active low, we must invert the input variables. Since we are assuming double-rail inputs, input inverters are not shown in Figure 3.11. The final output of the circuit is already active high; so no conversion is needed there. The final connection is that between the first and second OR form. Since the output of the first OR form is active high and the input of the second OR form is active low, a direct connection cannot be made. To match the levels, a NAND gate used as an inverter is inserted. Since the inverter is converting from active high to active low, we show the AND form of the NAND gate as an inverter since it has active high inputs and an active low input.

3.4.2 Logic Diagram Design Using NOR Gates

As with the NAND gates, the first step in implementing a circuit using only NOR gates is replacing all OR gates with the OR form of the NOR gate and all AND gates with the AND form of the NOR gate. Doing this for Figure 3.5 yields Figure 3.12.

First we must provide active high inputs and outputs to the real world. The AND form of the NOR gate has active low inputs; so the input variables must be complemented. Since we are assuming double-rail inputs, the input inverters are not shown in Figure 3.13. The final output of the circuit is also active low; so an inverter must be added to convert this output to active high. A NOR gate is used as an inverter, and to show the active low input and active high output required of the inverter, the AND form of the NOR gate is shown. Finally, connections can be made directly between the outputs of the AND forms and the inputs of the OR form since both are active high. The final logic diagram using only NOR gates is shown in Figure 3.13.

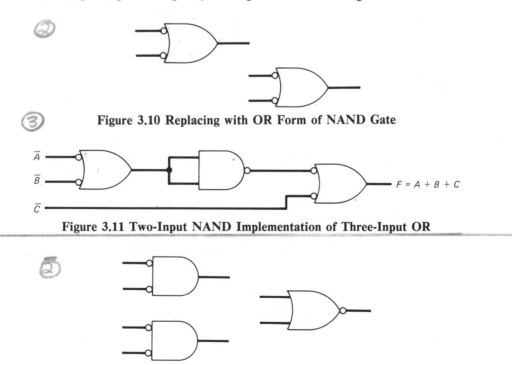

Figure 3.10 Replacing with OR Form of NAND Gate

$F = A + B + C$

Figure 3.11 Two-Input NAND Implementation of Three-Input OR

Figure 3.12 Replacing with AND and OR Form of NOR Gate

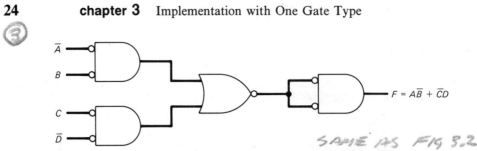

$$F = A\bar{B} + \bar{C}D$$

SAME AS FIG 3.2

Figure 3.13 NOR Implementation of Equation 3.1

Lab Exercise

Objective

This lab allows the student to apply concepts from Chapter 3 to implement a network that he or she must design. First, given a description of the digital network desired, the student must define in truth table, SOP and POS form, the function of this network. From these definitions, the topics covered in Chapter 3 are used to implement the SOP and POS expressions using NAND gates only.

Procedure

1. Design and build an even parity generator for a 3-bit word. This circuit will have three inputs (the 3 bits of the word) and one output. This output should be 1 if there is an odd number of 1s in the input word; otherwise, the output should be 0 (if there is an even number of 1s in the input word). Produce a design using gate equivalency rules containing:
   (a) A truth table defining the function as previously described.
   (b) An SOP equation for the truth table.
   (c) A POS equation for the truth table.
   (d) A logic diagram to implement the SOP equation using NAND gates only. (Show AND and OR forms.)
   (e) A logic diagram to implement the POS equation using NAND gates only. (Show AND and OR forms.)
   Wire (d) and (e) and verify their operation.

   Have the lab instructor inspect your work after each circuit is working.

Review Questions

3.1 Given the following boolean expression:

$$F = \bar{A}B\bar{C} + \bar{A}BC + A\bar{B}C$$

produce using algebraic manipulation:
   (a) An equivalent expression using only NAND operations. (Show all work and intermediate equations.)
   (b) A logic diagram implementing the preceding equation using only NAND gates.
   (c) An equivalent expression using only NOR operations. (Show all work and intermediate equations.)
   (d) A logic diagram implementing the preceding equation using only NOR gates.

3.2 Repeat Question 3.1 for the following expression:

$$F = \bar{A}\bar{B}\bar{C} + \bar{A}\bar{B}C + A\bar{B}C + AB\bar{C} + ABC$$

3.3 Given the equation in Question 3.1, produce using gate equivalency rules:
   (a) An original logic diagram using AND and OR gates.

(b) A logic diagram using only NAND gates. (Show AND and OR forms.)

(c) A logic diagram using only NOR gates. (Show AND and OR forms.)

3.4 Repeat Question 3.3 using the expression in Question 3.2.

3.5 Implement a three-input AND gate using only two-input NAND gates. Use gate equivalency rules to design the logic diagram.

3.6 Repeat Question 3.5 to implement a three-input OR gate.

3.7 Repeat Question 3.5 to implement a three-input NAND gate.

3.8 Using gate equivalency rules, implement the following expression using only NAND gates:

$$F = (A + \bar{B}) \cdot (CD + E)$$

3.9 Using gate equivalency rules, implement the expression in Question 3.8 using only NOR gates.

3.10 Given the following expression:

$$F = (A + \bar{B})C + \bar{A}B\bar{C}$$

(a) Implement using AND and OR gates.

(b) Implement using only NAND gates.

3.11 What is the advantage of using NAND logic over NOR logic to implement a function with TTL components?

3.12 Using gate equivalency rules, design a logic diagram to implement the function defined by the accompanying truth table using only NAND gates and inverters.

A	B	C	F
0	0	0	0
0	0	1	0
0	1	0	1
0	1	1	1
1	0	0	0
1	0	1	1
1	1	0	0
1	1	1	1

3.13 Using gate equivalency rules, design a logic diagram to implement the function defined by the accompanying truth table using only NOR gates and inverters.

A	B	C	F
0	0	0	0
0	0	1	0
0	1	0	0
0	1	1	1
1	0	0	1
1	0	1	1
1	1	0	0
1	1	1	0

4

Expression Reduction Techniques

Objective

An introduction to methods for reducing boolean expressions to simplest form. Algebraic manipulation and Karnaugh maps are demonstrated.

4.1 Expression Reduction

Techniques studied in Chapter 3 reduced hardware requirements by implementing expressions using just one gate type. Further reduction is possible by simplifying the original equation before attempting to implement it. This reduction is typically performed algebraically or with Karnaugh maps.

4.2 Algebraic Reduction

In Chapter 2, we converted an expression to normal form by multiplying by the missing variables and their complements. If we then removed these combinations from the normal-form equation, the simpler original equation would result. This is the basic operation in expression reduction—removing combinations of a variable and its complement. Consider the following equation, already in simplest form:

$$F = BC + A\overline{B}$$

This is converted into normal form by multiplying by the missing variables and their complements:

$$F = (A + \overline{A})BC + A\overline{B}(C + \overline{C})$$
$$F = ABC + \overline{A}BC + A\overline{B}C + A\overline{B}\overline{C} \qquad (4.1)$$

If given this final equation in normal form and asked to reduce it to simplest form, how would you do this? Since we derived it from the original equation, the process is obvious. The first two terms in the normal-form equation were generated by:

$$ABC + \overline{A}BC = (A + \overline{A})BC$$

and the last two by:

$$A\overline{B}C + A\overline{B}\overline{C} = A\overline{B}(C + \overline{C})$$

Since any variable ORed with its complement is 1, these combinations can be removed yielding the original equation:

$$F = BC + A\overline{B}$$

The process employed here is simple. If in any two terms one variable appears in both its true and complemented form and the remaining variables are the same in both terms, then the variable and its complement can be removed, and the result is simply one term composed of the remaining variables. For example, in the previous normal-form equation, the first two terms contain A in both true and complemented form, and the remaining variables are the same in both terms (B and C). Thus A and its complement can be removed, resulting in one term, BC, as previously demonstrated.

To further illustrate this process, we follow the steps to simplify the following expression:

$$F = \overline{A}\overline{B}C + \overline{A}BC + A\overline{B}C + ABC \qquad (4.2)$$

In the first two terms, B is present in both its true and complemented form. In addition, all remaining variables are the same in both terms (A and C). Thus the first two terms can be reduced to:

$$\overline{A}\overline{B}C + \overline{A}BC = \overline{A}(B + \overline{B})C = \overline{A}C$$

In the last two terms, B is again present in both its true and complemented form, and again all remaining variables are the same in both terms (A and C). Thus the last two terms can be reduced to:

$$A\overline{B}C + ABC = A(B + \overline{B})C = AC$$

yielding a reduced expression:

$$F = \overline{A}C + AC$$

Looking at this equation, we see additional reduction is possible. Since A appears in both its true and complemented form and the remaining variables are the same in both terms (C), the expression can be reduced to:

$$\overline{A}C + AC = (A + \overline{A})C = C$$
$$F = C$$

Therefore, the hardware required to implement this function is merely a wire running from input C to the output F!

The simplification possible and the resulting reduction in hardware requirements is obvious. But algebraic reduction is often difficult—especially on complex expressions. The proper sequence of operations that must be performed is not always clear. But just as a graphical method made implementing circuits with just one gate type in Chapter 3 simpler, so a graphical method can be used to make expression reduction simpler.

4.3 Karnaugh Maps

Karnaugh maps or K-maps provide a graphical method for reducing expressions to minimum form. A *K-map* is a specially arranged truth table such that possible reductions are made more obvious. Shown in Figures 4.1, 4.2, and 4.3 are the forms of (a) 2-input, (b) 3-input, and (c) 4-input K-maps. Instead of listing all the possible input combinations as in a standard truth table, K-maps use the positions of the squares in the map to represent the different input combinations.

To aid in determining the input combination represented by a square, the rows and columns are labeled with the partial input combination represented by that row or column. For example, what input combination is represented by the square in row 2, column 4, of the 4-input K-map? Row 2 is labeled 01; column 4 is labeled 10. Caten-

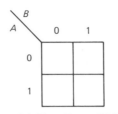

Figure 4.1 Two-Input K-Map

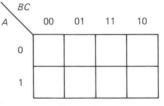

Figure 4.2 Three-Input K-Map

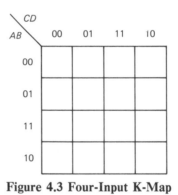

Figure 4.3 Four-Input K-Map

ating these labels in row-column order yields 0110, which is minterm 6 or $\overline{A}BC\overline{D}$. The square in row 3, column 2, represents the input combination 1101, which is minterm 13 or $AB\overline{C}D$.

A more in-depth observation reveals that the label 00 on row 1 indicates that A and B are both 0 for that entire row. Likewise, the label 10 on column 4 indicates that C is 1 and D is 0 for the entire column.

To represent a function with a K-map, 1s and 0s are placed in the proper squares based on the input combination represented by each square.

The special arrangement of the K-map insures that the input combinations represented by two adjacent squares are identical except that one variable appears in its true form in one square and in its complemented form in the other. Note that this is simply the criterion specified in Section 4.1 for the reduction of two terms. Therefore, any two adjacent squares are eligible for reduction.

This relation can be seen by looking at any two adjacent squares in a K-map. For example, consider the first two squares in row 1 of the 4-input K-map. The first square represents the input combination 0000, or $\overline{A}\,\overline{B}\,\overline{C}\,\overline{D}$. The second square represents the input combination 0001, or $\overline{A}\,\overline{B}\,\overline{C}D$. Between these two terms, all variables are identical except that D appears in its true form in the second square and in its complemented form in the first square.

Note that this relation holds for any two adjacent squares whether paired vertically or horizontally. In addition, the relation even holds between outside edges (e.g., row 1, squares 1 and 4; or column 1, rows 1 and 4, and so on); thus outside edges are considered *adjacent*.

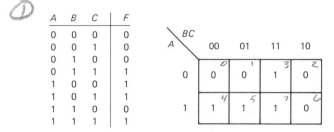

A	B	C	F
0	0	0	0
0	0	1	0
0	1	0	0
0	1	1	1
1	0	0	1
1	0	1	1
1	1	0	0
1	1	1	1

Figure 4.4 Truth Table and K-Map for Equation 4.1

4.3.1 Using Karnaugh Maps

To use a K-map for reduction, first fill in the K-map with the output of the function. The standard truth table for Equation 4.1 is shown in Figure 4.4. The output of this function is 1 for the input combination 011. This is represented by placing a 1 in row 1 (0), column 3 (11), of the K-map. The output is also 1 for minterm 4 (100). This is represented by placing a 1 in row 2 (1), column 1 (00), of the K-map. Repeating this for the remaining terms results in the K-map shown in Figure 4.4.

Since adjacent squares represent terms that can be reduced, look for and circle groups of adjacent 1s. A group must consist of 1, 2, 4, 8, or 16 1s—that is, a power of 2. In general, the larger the group circled, the simpler the reduced expression will be; so first look for groups of 16 (not applicable in this example, of course), then groups of 8, then groups of 4, then groups of 2, and finally, single 1s, until all 1s have been grouped. There are two groups of 1s in the K-map shown in Figure 4.4; these are circled in Figure 4.5.

The resulting SOP expression will consist of one product term for each group circled. Therefore, the fewer groups circled, the simpler the expression. In the K-map in Figure 4.5, two groups are circled. The resulting product term for each group is composed of those variables that are the same for the entire group. For example, in the vertical group, both B and C are 1 for the entire group. A is 0 for the upper square and 1 for the lower square. Therefore, only B and C remain the same, and the resulting product term is BC. In the horizontal group, A is 1 in both squares, and B is 0 in both squares. C is 0 in the left square and 1 in the right square. Therefore, the resulting product term is $A\overline{B}$. Combining these two product terms results in a final reduced expression:

$$F = BC + A\overline{B}$$

As a second example, we reduce Equation 4.2 using K-maps. Shown in Figure 4.6 is the standard truth table for Equation 4.2 and the resulting K-map. Looking first for groups of 8, we find there are none. Looking for groups of 4, we find one as shown in Figure 4.6. Since all 1s have been circled, there is no need to look for other groups. The resulting SOP equation will have only one product term since there is only one group circled. That product term consists of those variables that are the same for the entire group. Looking at Figure 4.6 we see A is 0 for the upper two squares and 1 for the lower two squares. In addition, B is 0 for the left two squares and 1 for the right

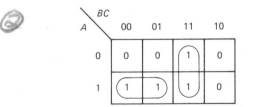

Figure 4.5 Circling Groups of Adjacent 1s

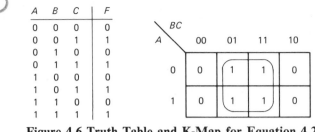

A	B	C	F
0	0	0	0
0	0	1	1
0	1	0	0
0	1	1	1
1	0	0	0
1	0	1	1
1	1	0	0
1	1	1	1

Figure 4.6 Truth Table and K-Map for Equation 4.2

two squares. Only C is the same (1) for all four squares. Therefore, the resulting product term is simply C, and the final reduced expression for F is:

$$F = C$$

As a final example, we look at a slightly more complex case. Shown in Figure 4.7 is the truth table and corresponding K-map for a four-input network.

Figure 4.7 illustrates how this function is reduced. As can be seen, a group of four has been formed using the fact that outside edges are considered adjacent. Forming this larger group of four rather than two groups of two results in a simpler final expression. Note a second application of wraparound in forming the group of two 1s between the top and bottom rows. Finally we note the group of two formed using a 1 that has already been circled. In general, if a larger group can be formed by including 1s that have already been circled, use them. As shown in Figure 4.7, using a 1 that had already been circled resulted in a group of two 1s rather than just a single 1.

The resulting expression will consist of three product terms since there are three groups circled on the K-map. Looking first at the group of four, we find B and D remain constant: B is always 1; D is always 0. This yields the product term $B\overline{D}$. For the vertical group of two, B is 0, C is 0, and D is 1. This results in a product term for this group of $\overline{B}\,\overline{C}D$. In the final group of two, A is 0, B is 1, and C is 1. This yields the product term $\overline{A}BC$. ORing these product terms results in the final reduced expression:

$$F = B\overline{D} + \overline{B}\,\overline{C}D + \overline{A}BC$$

4.3.2 Don't Care Conditions

Many functions have several input combinations for which no particular output is defined. These are called *don't-care* conditions. In a K-map, a don't-care condition can be used as 0 or 1—whichever results in a simpler expression.

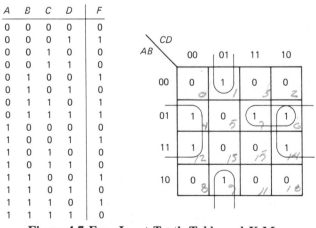

A	B	C	D	F
0	0	0	0	0
0	0	0	1	1
0	0	1	0	0
0	0	1	1	0
0	1	0	0	1
0	1	0	1	0
0	1	1	0	1
0	1	1	1	1
1	0	0	0	0
1	0	0	1	1
1	0	1	0	0
1	0	1	1	0
1	1	0	0	1
1	1	0	1	0
1	1	1	0	1
1	1	1	1	0

Figure 4.7 Four-Input Truth Table and K-Map

Table 4.1 "Greater than 5" Truth Table

A	B	C	D	F
0	0	0	0	0
0	0	0	1	0
0	0	1	0	0
0	0	1	1	0
0	1	0	0	0
0	1	0	1	0
0	1	1	0	1
0	1	1	1	1
1	0	0	0	1
1	0	0	1	1
1	0	1	0	x
1	0	1	1	x
1	1	0	0	x
1	1	0	1	x
1	1	1	0	x
1	1	1	1	x

For example, consider the network defined by the truth table shown in Table 4.1. This network accepts a 4-bit BCD digit as an input. The output is 1 if the digit is greater than 5; otherwise, the output is 0. Since it is BCD, valid binary inputs are only the digits 0 thru 9. Therefore, minterms 10 thru 15 are considered don't-care conditions and are marked with an x. The K-map for this truth table is shown in Figure 4.8. In this K-map we see two groups of 1s as circled in Figure 4.5. The group in the bottom left corner has A, $\overline{B}$, and $\overline{C}$ in common for the entire group, resulting in the product term $A\overline{B}\overline{C}$. The other group has $\overline{A}$, B, and C in common resulting in the product term $\overline{A}BC$. Combining these two terms yields a final reduced expression:

$$F = A\overline{B}\,\overline{C} + \overline{A}BC$$

But, since the Xs represent don't-care conditions, we can assume they are either 0s or 1s. Since assuming they were 1s would enable us to form larger groups, we can circle groups as shown in Figure 4.9. Now we have a group of eight and a group of four. The only variable common to all terms in the group of eight is A; thus A is the first product term. The variables common to the group of four are B and C; thus BC is the second product term, yielding a simpler equation:

$$F = A + BC$$

The use of don't cares will typically enable further reduction of an expression as demonstrated here.

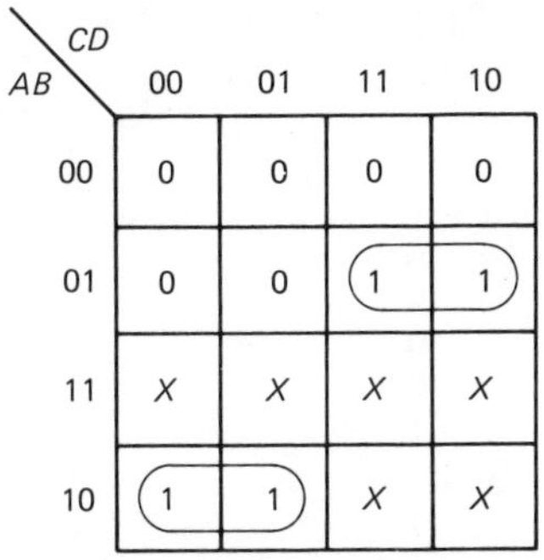

Figure 4.8 K-Map for Table 4.1

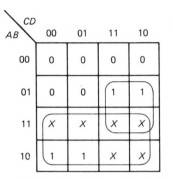

Figure 4.9 K-Map Using Don't Cares

Lab Exercise

Objective

This lab requires that the student use K-maps to reduce the expressions obtained from a simple design that the student must perform. In addition, the implementation must use only NAND gates and inverters, requiring application of techniques learned in Chapter 3.

Procedure

1. Design a circuit that given a 4-bit *BCD* digit (0–9) will produce a 2-bit quotient and 2-bit remainder for a division by 3. This circuit will have four inputs—the 4 bits of the *BCD* digit— and four outputs—2 bits for the quotient and 2 bits for the remainder.
   (a) Produce a truth table illustrating the behavior of this circuit. This truth table will have four inputs and four outputs.
   (b) Write the SOP equation for each of the four outputs of this circuit.
   (c) Using K-maps (and don't cares!), reduce each of the output equations to simplest form. Show K-maps and final equations.
   (d) Produce a logic diagram for implementing each of the reduced equations using NAND gates and inverters only.
   (e) Wire and test your circuit. This should be performed in two stages. First wire and test the quotient functions; then finish by wiring the remainder functions.

   Have the lab instructor check your work after each wiring stage is completed.

Review Questions

4.1 Reduce the following expressions using algebraic reduction. Show all intermediate equations.
   (a) $F(A, B, C) = \overline{A}\,\overline{B}\,\overline{C} + \overline{A}\,\overline{B}C + AB\overline{C} + ABC$
   (b) $F(A, B, C) = A\overline{B}\,\overline{C} + \overline{A}\,\overline{B}C + AB\overline{C} + \overline{A}BC$

4.2 Reduce the following expressions using K-maps. Show the K-map used.
   (a) $F(A, B, C) = ABC + AB\overline{C} + A\overline{B}C + \overline{A}BC + \overline{A}\,\overline{B}C$
   (b) $F(A, B, C, D) = \overline{A}\,\overline{B}CD + \overline{A}BCD + \overline{A}BC\overline{D} + \overline{A}B\overline{C}\,\overline{D} + \overline{A}BCD + AB\overline{C}\,\overline{D}$
   (c) $F(A, B) = \overline{A}\,\overline{B} + A\overline{B}$
   (d) $F(A, B, C) = \overline{A}\,\overline{B}\,\overline{C} + \overline{A}BC + A\overline{B}C + AB\overline{C}$

4.3 Using K-maps, derive the simplest expression for F as defined by the accompanying truth table.

A	B	C	F
0	0	0	0
0	0	1	1
0	1	0	1
0	1	1	0
1	0	0	1
1	0	1	0
1	1	0	0
1	1	1	1

4.4 Using K-maps, derive the simplest expression for F as defined by the accompanying truth table.

A	B	C	F
0	0	0	1
0	0	1	0
0	1	0	1
0	1	1	1
1	0	0	1
1	0	1	0
1	1	0	0
1	1	1	1

4.5 Given the following truth table, derive the simplest expression for F using K-maps.

A	B	C	D	F
0	0	0	0	1
0	0	0	1	1
0	0	1	0	0
0	0	1	1	1
0	1	0	0	1
0	1	0	1	0
0	1	1	0	1
0	1	1	1	1
1	0	0	0	0
1	0	0	1	0
1	0	1	0	1
1	0	1	1	0
1	1	0	0	1
1	1	0	1	1
1	1	1	0	0
1	1	1	1	1

4.6 Given the following truth table, derive the simplest expression for F using K-maps.

A	B	C	D	F
0	0	0	0	1
0	0	0	1	1
0	0	1	0	0
0	0	1	1	1
0	1	0	0	1
0	1	0	1	0
0	1	1	0	1
0	1	1	1	1
1	0	0	0	0
1	0	0	1	0
1	0	1	0	1
1	0	1	1	0
1	1	0	0	1
1	1	0	1	1
1	1	1	0	0
1	1	1	1	1

4.7 Expression (a) is a simplified version of expression (b). Find the don't cares, if any. (*Hint:* Use a K-map.)

(a) $F(A, B, C) = A\overline{B} + C$

(b) $F(A, B, C) = \overline{A}\,\overline{B}C + A\overline{B}C + A\overline{B}\,\overline{C}$

4.8 Expression (a) is a simplified version of expression (b). Find the don't cares, if any. (*Hint:* Use a K-map.)

(a) $F(W, X, Y, Z) = \overline{W}Z + XZ$

(b) $F(W, X, Y, Z) = \overline{W}\,\overline{X}\,\overline{Y}Z + \overline{W}\,\overline{X}YZ + \overline{W}X\overline{Y}Z + WX\overline{Y}Z$

4.9 Simplify the following expression:

$$F(A, B, C, D) = (\overline{A} + \overline{B} + \overline{D})\cdot(A + \overline{B} + \overline{C})\,(\overline{A} + B + \overline{D})\,(B + \overline{C} + \overline{D})$$

4.10 Simplify the following expression with the accompanying don't cares:

$$F(A, B, C, D) = \overline{A}\,\overline{B}D + \overline{A}CD + \overline{A}BC$$
$$\text{don't care} = \overline{A}B\overline{C}D + ACD + A\overline{B}\,\overline{D}$$

4.11 You are to design a combinational circuit that when given a 3-bit binary number as an input will produce a binary number equal to the square of the number as an output. The truth table for such a circuit follows. Find the simplest expressions for the output quantities in SOP form.

Inputs			Outputs					
x	y	z	A	B	C	D	E	F
0	0	0	0	0	0	0	0	0
0	0	1	0	0	0	0	0	1
0	1	0	0	0	0	1	0	0
0	1	1	0	0	1	0	0	1
1	0	0	0	1	0	0	0	0
1	0	1	0	1	1	0	0	1
1	1	0	1	0	0	1	0	0
1	1	1	1	1	0	0	0	1

32   16   8   4   2   0

5

Multiplexors and Demultiplexors

P241

Objectives

1. An introduction to the function and use of multiplexors and demultiplexors in digital circuits.
2. An in-depth study of the operation of multiplexors and demultiplexors at the gate level.

5.1 Multiplexors and Demultiplexors Defined

Our study of the basics of digital logic and circuit design provide the background needed to understand and design some of the important combinational networks used in digital circuits. In this chapter, we study two of these important networks, the multiplexor and the demultiplexor.

The multiplexor (mux), provides the function of a rotary switch, selecting one of several inputs to connect to a single output. The multiplexor is often referred to as a *selector*. The demultiplexor (demux), provides the inverse function, connecting a single input to one of several outputs. The demultiplexor is often referred to as a *decoder*. These two functions are used extensively in digital networks and are capable of performing in several different roles.

Used in tandem, a mux-demux combination can be used for serial communication to reduce the number of wires required to pass data. The pair can be used in a similar manner to run multidigit displays such as those found in calculators. Muxes are often used in digital circuits to control signal and data routing. For example, a multiplexor can be used to select the input to a particular register from one of several sources. Decoders are often used in computers to provide address decoding. Based on certain address lines, the decoder can provide enable signals to the proper memory chips. In addition, both muxes and demuxes can be used to evaluate simple boolean expressions using less hardware than if individual logic gates were used.

There are many additional applications; these applications just begin to illustrate the flexibility of these two functions.

5.2 The Multiplexor

As previously mentioned, a mux acts like a rotary switch connecting one of several inputs to a single output. The selection of which input to connect to the output is determined by additional inputs called select or control lines. The input selected is determined by the binary equivalent of the value placed on the select lines. For example, consider a mux that selects one of four inputs to connect to the output. This is referred

35

Table 5.1 4-to-1 Mux Truth Table

S2	S1	F
0	0	D0
0	1	D1
1	0	D2
1	1	D3

to as a 4-to-1 mux. To select one of four inputs, there must be four unique combinations of the select lines. This requires two select lines providing the four unique combinations: 00, 01, 10, and 11. A select-line combination of 00 would select input 0, select-line combination 01 would select input 1, and so on.

The function of the multiplexor is illustrated in the truth table shown in Table 5.1. This truth table shows the output F as a function of the select-line inputs. Instead of listing all possible states of the data inputs, this simplified form of a truth table shows the output as the data from input line 0 ($D0$), or input line 1 ($D1$), and so on. As can be seen from Table 5.1, we want to pass $D0$ when $S2$ and $S1$ are both 0. Likewise, we want to pass $D1$ when $S2$ is 0 and $S1$ is 1, and so on for the remaining input combinations. This is implemented by the following expression:

$$F = \overline{S2}\cdot\overline{S1}\cdot D0 + \overline{S2}\cdot S1\cdot D1 + S2\cdot\overline{S1}\cdot D2 + S2\cdot S1\cdot D3$$

Shown in Figure 5.1 is a NAND-gate implementation of a 4-to-1 multiplexor. To better illustrate the function of the multiplexor as a complete "unit," all variable complements have been generated internally.

Since the multiplexor function is so useful, many TTL chips exist that perform the equivalent operation of the circuit shown in Figure 5.1. For example, the 74153 contains two 4-to-1 multiplexors, the 74151 contains one 8-to-1 multiplexor, and the 74157 contains four 2-to-1 multiplexors.

A multiplexor is typically shown in a circuit as a single functional unit, not as

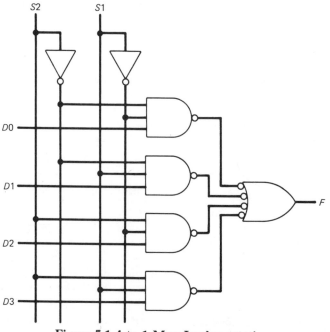

Figure 5.1 4-to-1 Mux Implementation

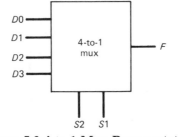

Figure 5.2 4-to-1 Mux Representation

the gates comprising the circuit. A typical representation of a 4-to-1 multiplexor is shown in Figure 5.2.

5.3 The Demultiplexor

The demultiplexor connects one input to one of several outputs. The output is selected via select lines as it is with the mux. The most common use of a demultiplexor is as a decoder. In fact, demuxes are typically referred to as decoders. A decoder sets the output line selected by the select lines to 0. This function is easily implemented with the demux by tying the data input line to 0.

To illustrate the behavior of a demux, consider the operation of a demux that connects one input to one of four outputs. This is referred to as a 1-to-4 demultiplexor. Used as a decoder, it is referred to as a 2-to-4 decoder since two select lines select one of four outputs. The operation of this demux is illustrated in Table 5.2. This truth table shows the four outputs of the demux as a function of the data in DI and select lines.

Notice that the *default* state of an output is 1. Therefore, any outputs not selected will be 1. Thus, as Table 5.2 illustrates, when the data in DI input is 1, all outputs will be 1 regardless of the select-line combination.

A decoder sets the output selected by the select-line inputs to 0. As can be seen in Table 5.2, this is accomplished by tying the data input to 0. Thus the data input can be thought of as an active-low enable for the decoder. If this enable is not low, the decoder will not function (all outputs will always be 1). Often decoders will have more than one enable line, requiring all enable lines to be properly enabled before the decoder functions.

The functions illustrated in Table 5.2 can be expressed algebraically as:

$$F0 = \overline{\overline{DI}\ \overline{S2}\ \overline{S1}}$$
$$F1 = \overline{\overline{DI}\ \overline{S2}\ S1}$$
$$F2 = \overline{\overline{DI}\ S2\ \overline{S1}}$$
$$F3 = \overline{\overline{DI}\ S2\ S1}$$

Shown in Figure 5.3 is a NAND-gate implementation of the 1-to-4 demux (2-to-4 decoder).

As with multiplexors, the demultiplexor circuit is available in several variations

Table 5.2 1-to-4 Demux Truth Table

DI	$S2$	$S1$	$F0$	$F1$	$F2$	$F3$
0	0	0	0	1	1	1
0	0	1	1	0	1	1
0	1	0	1	1	0	1
0	1	1	1	1	1	0
1	x	x	1	1	1	1

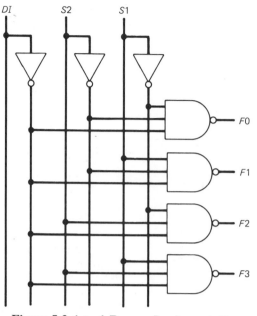

Figure 5.3 1-to-4 Demux Implementation

on standard TTL chips. For example, the 74138 contains one 1-to-8 demux (3-to-8 decoder), and the 74139 contains two 1-to-4 demuxes (2-to-4 decoders).

As with the multiplexor, a demux is typically shown in a circuit as a separate functional unit, not as the gates comprising the demux. Shown in Figure 5.4 are typical demux and decoder representations of the circuit in Figure 5.3. Note the presence of active-low indicators at the enable input and each of the outputs on the decoder representation. This indicates the decoder is enabled with a low input, and an output is set to low when selected.

5.4 Multiplexor-Demultiplexor Applications

Now that the basic operation of the mux and demux is understood, we look in detail at the applications mentioned in Section 5.1.

5.4.1 Communication

Consider a situation in which two 16-bit computers must communicate. This communication could easily be accomplished using parallel transmission of all 16 bits. This would require at least 18 conductors, one for each bit, a ground reference, plus an additional data strobe line to indicate that a word of data is available.

An alternate method is illustrated in Figure 5.5. Using a multiplexor to select a single bit of the input word and a demultiplexor to connect that data to the proper bit

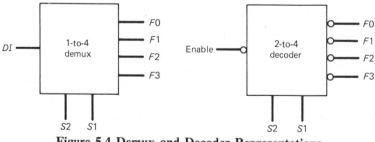

Figure 5.4 Demux and Decoder Representations

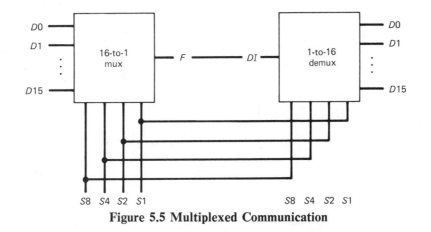

Figure 5.5 Multiplexed Communication

of the output word, serial communication can be established using just six lines: one for the data line, four select lines, and a ground reference. An entire word can be transmitted by sequencing the select lines from 0 thru 15. This technique is termed *multiplexing.*

Though fewer conductors are required using multiplexing, the parallel transfer method is inherently capable of faster transmission rates. But in many applications, the speed attainable with multiplexing is sufficient and the reduction in connections is to great advantage.

For example, the TI 9900 series of 16-bit microprocessors uses a scheme similar to that shown in Figure 5.5 to send and receive data from I/O controller chips. Since data is sent and received serially, the I/O chips can be bit addressed, sending or receiving anywhere from 1 to 16 bits at a time. In addition, since there is no need for 8 or 16 data lines, the I/O chips are the smaller 16-, 18-, or 20-pin chips rather than the wide 24- or 40-pin chips. Thus in this application not only does a multiplexed scheme reduce connections but it results in increased flexibility by providing bit-addressable I/O.

5.4.2 Data Routing

In a typical computer, the program counter (PC) can be loaded from several different sources. For example, in straight-line execution, the PC is simply incremented after each instruction. In this case, the PC might be loaded from a counter. If an absolute-jump instruction is executed, the address specified by the jump instruction must be loaded into the PC. Often this address is present in the instruction register (IR). If a relative branch is executed, then the displacement specified must be added to the current PC. In this case, the PC might be loaded from the output of the arithmetic logic unit (ALU).

A mux is ideally suited to this sort of application—selecting one of multiple input sources. In this situation, the PC is connected to the output of the mux, and the inputs to the mux are the counter, instruction register, and ALU as just described. By placing different values on the select lines, the input for the PC can be selected from one of these sources. The value placed on the select lines might come from a field in a microinstruction or from a hardwired sequencer controlling the execution of instructions.

There is one obvious problem with this approach. Assume the computer we are working with uses 16-bit addresses. Thus the mux needs to select between inputs of 16 bits each. A mux only has one input for each select line combination, not 16. How can a full 16-bit value be routed using a mux? Figure 5.6 illustrates how this can be accomplished by using a series of muxes, one for each bit of the word. In this figure,

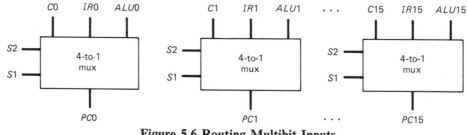

Figure 5.6 Routing Multibit Inputs

$C0$ refers to bit 0 of the counter, $IR0$ refers to bit 0 of the address in the instruction register, $ALU0$ refers to bit 0 of the ALU output, and so on for the remaining 15 bits of the address. The same select lines run to each multiplexor. Since there are only three inputs in this case, the fourth input on the multiplexors is not used.

5.4.3 Address Decoding

Consider a simple microprocessor system with 256 bytes of memory. Assume this memory is made up of one 256×8 RAM chip (256 8-bit registers). To address the locations in this chip, 8 address lines are required (addresses 00–FF). This configuration is shown in Figure 5.7. Note that the RAM chip has an active-low enable input. The RAM is enabled in this case by tying the enable input low. The 8 data lines are not shown so that the complexity of the diagram can be reduced.

Now we want to increase the RAM of the computer to 1K. This obviously requires four 256×8 RAM chips. The addressing of these chips breaks down as follows:

Address	RAM Chip
0000–00FF	0
0100–01FF	1
0200–02FF	2
0300–03FF	3

The lower 8 address lines ($A7$–$A0$) still run to the address lines on each of the four chips to select one of the 256 locations in each RAM chip. The next two address lines ($A8$ and $A9$) determine which of the four RAM chips to access.

When $A9$ and $A8$ are both 0, RAM chip 0 should be enabled. When $A9$ is 0 and

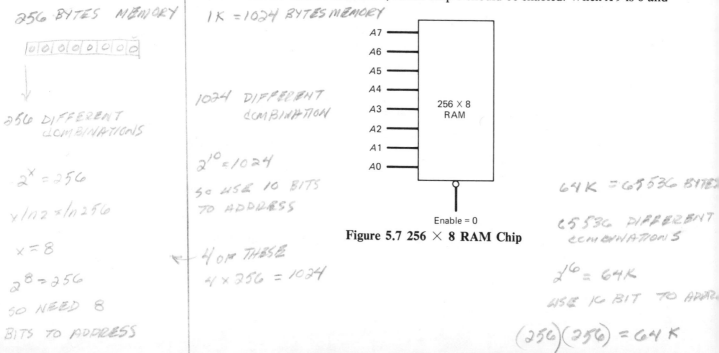

Figure 5.7 256×8 RAM Chip

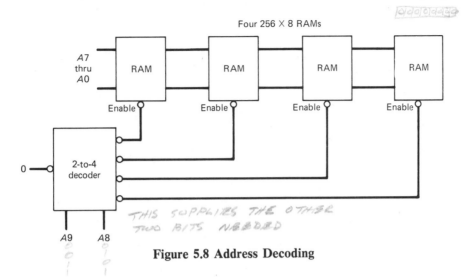

Figure 5.8 Address Decoding

$A8$ is 1, RAM chip 1 should be enabled, and so on. Using address lines $A9$ and $A8$ as select-line inputs on a 2-to-4 decoder, we cause one of the four outputs to go low based on the two address lines. This output is connected to the enable input on the RAM chip to be enabled by that particular address-line combination. This arrangement is shown in Figure 5.8.

5.4.4 Expression Implementation Using Multiplexors

Multiplexors can be used to easily implement a truth table without complex AND-OR circuits. Consider the truth table shown in Table 5.3. A digital network to implement this truth table should have an output of 1 for minterm 0, an output of 1 for minterm 1, an output of 0 for minterm 2, and so on. Using A, B, and C as select-line inputs on an 8-to-1 multiplexor, the output of the multiplexor is input 0 ($I0$) for minterm 0, input 1 ($I1$) for minterm 1, and so on. Thus the multiplexor can be used to implement this truth table by simply connecting the function inputs to the select lines and wiring the multiplexor inputs to 1 or 0 depending on the output for that minterm. An implementation of Table 5.3 using an 8-to-1 mux is shown in Figure 5.9.

In this example, we implemented a truth table with eight entries using an 8-to-1 mux. The function can also be implemented using a 4-to-1 mux. In general, a truth table with N entries can be implemented with either an N-to-1 mux or an $(N/2)$-to-1 mux. To implement with a 4-to-1 mux, function inputs A and B are connected to the two select lines on the mux. Viewing the truth table shown in Table 5.4 makes the connections required more obvious. For $AB = 00$, the output of the function is always 1; so input 0 ($I0$) on the mux is connected to 1. When $AB = 01$, the output of the function is 0 when C is 0, and 1 when C is 1. Therefore, we connect C to input 1 ($I1$) on the mux. When $AB = 10$, the output of the function is 1 when C is 0, and 0 when C is

Table 5.3

A	B	C	F
0	0	0	1
0	0	1	1
0	1	0	0
0	1	1	1
1	0	0	1
1	0	1	0
1	1	0	0
1	1	1	0

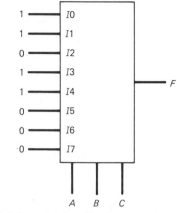

Figure 5.9 8-to-1 Mux Implementation of Table 5.3

Table 5.4

A	B	C	F
0	0	0	1
0	0	1	1
0	1	0	0
0	1	1	1
1	0	0	1
1	0	1	0
1	1	0	0
1	1	1	0

1. Thus we connect $\overline{C}$ to input 2 ($I2$) on the mux. For $AB = 11$, the output of the function is always 0. Therefore we simply connect input 3 ($I3$) to 0. This implementation is shown in Figure 5.10.

5.4.5 Expression Implementation Using Decoders

   Truth tables can also be implemented using decoders. A decoder will set the output selected by the select lines to 0. Consider a 3-to-8 decoder and a three-variable truth table. By placing the truth-table inputs on the select lines of the decoder, we set the output selected by the truth-table inputs to 0. For example, the output of the function defined in Table 5.3 is 1 for minterms 0, 1, 3, and 4. If any of these minterms is applied to the select lines on the decoder, that output will go low. If any of these outputs goes low, the output of the circuit should be 1. Thus we OR outputs 0, 1, 3, and 4. Since the decoder provides active-low outputs, the OR form of the NAND gate is used since it has active-low inputs. This implementation is shown in Figure 5.11.

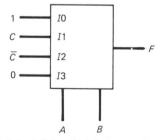

Figure 5.10 4-to-1 Mux Implementation of Table 5.3

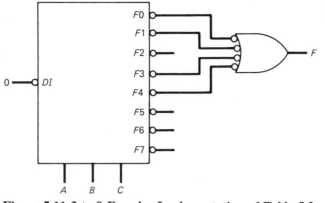

Figure 5.11 3-to-8 Decoder Implementation of Table 5.3

Lab Exercise

Objective

This lab provides experience using standard TTL multiplexors and demultiplexors. Proper implementation requires close attention to the details specified in the data sheets for each part. Use of the data sheets exposes the student to "real world" conditions.

Procedure

1. Design and wire circuits to implement the following expression using:
   (a) An 8-to-1 mux
   (b) A 4-to-1 mux
   (c) A 3-to-8 decoder

$$F = \overline{A}\,\overline{B}\,\overline{C} + \overline{A}B\overline{C} + \overline{A}BC + A\overline{B}C$$

2. A circuit is needed to display a 4-bit value on the LEDs of the logic designer. This value can come from two different sources. Design a circuit using four 2-to-1 muxes to select which of the two values to display on the LEDs. Use an input switch on the logic designer to select the input source. The two values to be displayed should be hardwired to the binary equivalent of 3 and 9. Wire and test your circuit.

3. Design a circuit that given a 2-bit value will light an LED corresponding to that value. Thus for each of the four possible input values (0 through 3), one of four LEDs should be lit. Obtain the input value from two switches on the logic designer. Wire and test your circuit.

Have the lab instructor inspect your work after each circuit is working.

Review Questions

5.1 Design an 8-to-1 mux using NAND gates and inverters.

5.2 Design a 1-to-8 demux using NAND gates and inverters.

5.3 Address decoding is needed for a 4K byte memory. This memory is built from the same 256 × 8 chips discussed in Section 5.4.3.
   (a) How many 256 × 8 RAM chips are needed?
   (b) What size decoder is needed to fully address all 4K of memory?
   (c) Assuming the least significant address bit is $A0$, which address lines must be connected to the select lines on the decoder?
   (d) Produce a logic diagram similar to Figure 5.8, illustrating your design for this 4K memory.

5.4 Design a circuit to implement the following truth table using:
   (a) An 8-to-1 mux

(b) A 4-to-1 mux

(c) A 3-to-8 decoder

A	B	C	F
0	0	0	0
0	0	1	1
0	1	0	1
0	1	1	0
1	0	0	1
1	0	1	0
1	1	0	0
1	1	1	1

5.5 Implement the following function using an 8-to-1 mux:

$$F = \sum m(4, 10, 13, 15)$$

5.6 Implement the following function using 2-to-1 multiplexors and a 4-to-1 multiplexor:

$$F(w, x, y, z) = \sum m(0, 1, 2, 4, 7, 11, 12, 13)$$

5.7 Implement the function of Question 5.6 using a single 8-to-1 mux.

5.8 Design a circuit to select an 8-bit word from one of two sources. Inputs to this circuit are the two 8-bit words and a select line to choose between the two words. The output of this circuit is the 8-bit value selected.

5.9 How many lines are required for 8-bit multiplexed communication using the technique described in Section 5.4.1? What are they?

5.10 The number of lines required for multiplexed communication could be reduced if the select lines could be generated independently at the transmit side and the receive side. That way the select lines would not have to be carried in the cable. What problems does this technique pose? How could these problems be avoided?

6

Binary Adders and Other Combinational Networks

Objectives

1. An in-depth study of the design and implementation of binary adders, including carry-lookahead techniques.
2. An overview of additional important combinational networks, including code converters, comparators, and display drivers.

6.1 Binary Adders

Adders play an important role in digital circuits—especially, of course, in computers. Adders used in digital circuits are typically binary adders. Multidigit binary values are added 1 bit position at a time, each position producing a sum digit and a carry out *(Cout)* into the next bit position. The process is analogous to the procedure you used in grade school to add multidigit decimal numbers, the only difference being the addition table used.

6.1.1 Half-Adders and Full-Adders

The addition table for two binary digits is shown in Table 6.1. This table shows both the sum and carry produced by the addition of 2 bits. Note that the sum is simply the EXCLUSIVE-OR function, and carry is simply the AND function. But the adder defined by Table 6.1 has one problem. Since the previous bit position can produce carry, the adder really needs three inputs: the 2 bits to be added, plus carry in from the previous bit position. An adder without the carry input is referred to as a *half-adder.* An adder that provides for the carry input *(Cin)* is termed a *full-adder.* Shown in Table 6.2 is the truth table for a full-adder.

A circuit to implement a full-adder is easily designed using techniques covered so far. Shown in Figure 6.1 are the reduced equations for the sum and carry outputs and the K-maps used to derive these equations. In these K-maps, the carry input has been abbreviated C.

A closer look at the truth table and these expressions reveals that the sum function is true whenever there is an odd number of 1s input, and the carry-out function is true whenever there is a majority of 1s input.

Table 6.1 Binary Addition Table

A	B	Sum	Carry
0	0	0	0
0	1	1	0
1	0	1	0
1	1	0	1

Table 6.2 Full Adder Truth Table

Cin	A	B	Sum	Cout
0	0	0	0	0
0	0	1	1	0
0	1	0	1	0
0	1	1	0	1
1	0	0	1	0
1	0	1	0	1
1	1	0	0	1
1	1	1	1	1

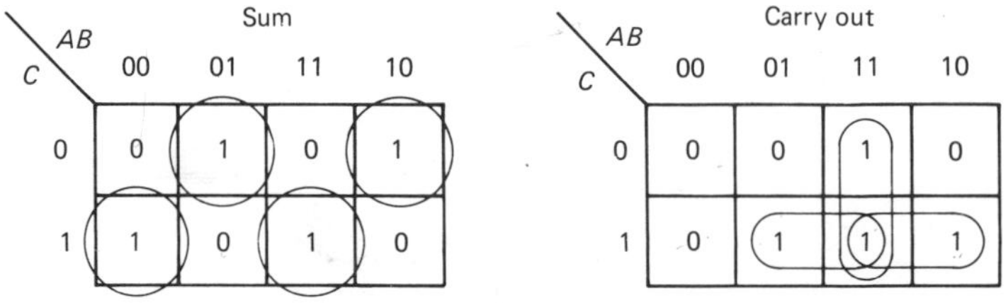

$Sum = \overline{A}\overline{B}C + \overline{A}B\overline{C} + ABC + A\overline{B}\overline{C}$

$Cout = AB + AC + BC$

Figure 6.1 Full-Adder Expression Reduction

Using these expressions for the sum and carry-out functions, the full-adder can be implemented using NAND gates and inverters as shown in Figure 6.2.

6.1.2 Multibit Adders

Using several full-adders, a multibit adder can be made. For example, a circuit to add two 4-bit values can be constructed from four full-adders as shown in Figure 6.3. Note the carry connections in the 4-bit adder. The carry input to each adder is the carry output from the previous adder. Since the output of each adder depends on the carry from the previous adder, each adder must "wait" for the carry output of the previous adder. This *carry ripple* slows down the addition circuit, especially as the number of bits in the adder increases.

Computing the time required to perform a multibit add will illustrate the effect of carry ripple. To calculate add time, the time delay associated with each full-adder must be determined. For simplicity, assume each gate shown in the full-adder circuit in Figure 6.2 has a switching delay of 10 ns. Thus the sum output is valid after three gate delays (inverter, NAND, NAND), or 30 ns. The carry output is valid after two gate delays (NAND, NAND), or 20 ns.

Using these delays, we see the carry out from bit 0 is valid after 20 ns. The carry out from bit 1 is valid 20 ns after that of bit 0, and so on. The final output of the 4-bit adder is not valid until the final-carry ($C3$) and final-sum ($S3$) bits are both valid. The final carry is valid 20 ns after $C2$; the final sum is valid 30 ns after $C2$. Therefore, the longer delay is incurred waiting for the final-sum output. Using these observations we can determine the total add time for a 4-bit adder as shown in the following equations:

$$Cin \text{ to } C0 = 20 \text{ ns}$$
$$C0 \text{ to } C1 = 20 \text{ ns}$$
$$C1 \text{ to } C2 = 20 \text{ ns}$$
$$C2 \text{ to } S3 = 30 \text{ ns}$$
$$\text{Add time} \quad 90 \text{ ns}$$

As the number of bits in the adder increases, the effect of the carry ripple is even more noticeable.

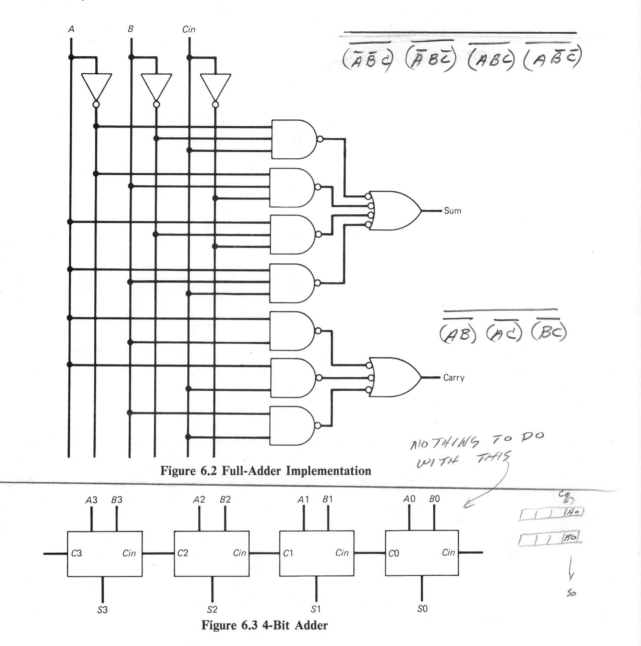

Figure 6.2 Full-Adder Implementation

Figure 6.3 4-Bit Adder

For example, in a 32-bit adder, add time would be:

32-bit add time $= 31 \times 20$ ns $+ 30$ ns $= 650$ ns

As can be seen, the add time increases proportionally to the number of bits to be added.

6.2 Carry Lookahead

The primary problem with the previous multibit adder is that the output of each adder depends on the carry output of the previous adder. If the carry values could be determined without having to ripple through all the previous stages, add time would be independent of the number bits in the adder and would therefore be reduced.

To achieve this, we note that a term (adder) will produce carry under two conditions. First, if both inputs are 1, carry will be generated regardless of the carry in. This is referred to as a *generate* condition. Second, if just one of the two inputs is a 1, carry will be produced only if there was carry in—that is, the previous term produced carry. This is referred to as a *propagate* condition, since the carry in is propagated to the carry out. These conditions for some term i can be illustrated in boolean algebra as:

$$Gi = AiBi \quad \text{(generate condition)}$$
$$Pi = Ai\overline{Bi} + \overline{Ai}Bi \quad \text{(propagate condition)}$$

To summarize, a term will produce carry if (1) it meets the generate condition (Gi), or (2) it meets the propagate condition (Pi) and there was carry in. This can be illustrated as:

$$Ci = Gi + (Pi)(C_{i-1})$$

Using this relation, an expression for each carry bit can be formed as illustrated here for a 4-bit adder:

$$C0 = G0 + P0 \cdot Cin$$

$$\begin{aligned} C1 &= G1 + P1 \cdot C0 \\ &= G1 + P1(G0 + P0 \cdot Cin) \\ &= G1 + P1 \cdot G0 + P1 \cdot P0 \cdot Cin \end{aligned}$$

$$\begin{aligned} C2 &= G2 + P2 \cdot C1 \\ &= G2 + P2(G1 + P1 \cdot G0 + P1 \cdot P0 \cdot Cin) \\ &= G2 + P2 \cdot G1 + P2 \cdot P1 \cdot G0 + P2 \cdot P1 \cdot P0 \cdot Cin \end{aligned}$$

$$\begin{aligned} C3 &= G3 + P3 \cdot C2 \\ &= G3 + P3(G2 + P2 \cdot G1 + P2 \cdot P1 \cdot G0 + P2 \cdot P1 \cdot P0 \cdot Cin) \\ &= G3 + P3 \cdot G2 + P3 \cdot P2 \cdot G1 + P3 \cdot P2 \cdot P1 \cdot G0 + P3 \cdot P2 \cdot P1 \cdot P0 \cdot Cin \end{aligned}$$

Looking at these equations, we see that each product term determines if carry could have come from a preceding term. This is possible only if a preceding term generated carry and all succeeding stages propagate it. For example, look at the final equation for $C3$. If term 0 generated carry, then for that carry to be output from term 3, stages 1, 2, and 3 must have all propagated that carry—hence the product term:

$$P3 \cdot P2 \cdot P1 \cdot G0$$

To determine the effectiveness of the carry-lookahead procedure, we need to determine the time required to calculate the carry terms shown previously. To simplify the logic diagram we note that the propagate condition is simply the EXCLUSIVE-OR function, which can be rewritten as:

$$Pi = Ai\overline{Bi} + \overline{Ai}Bi = \overline{AiBi + \overline{Ai}\,\overline{Bi}}$$

Using this equivalence, we can produce the generate (Gi) and propagate (Pi) terms used in the preceding equations as shown in Figure 6.4. As can be seen, two gate delays are associated with generating the propagate condition. This, in conjunction with the two-level nature of the carry expressions, results in a total of four gate delays, or 40 ns, to evaluate a carry term. Add times for a 4-bit and 32-bit adder using carry-lookahead logic can be calculated as shown.

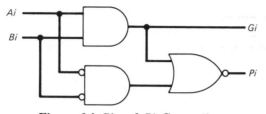

Figure 6.4 *Gi* and *Pi* Generation

Add time for a 4-bit adder:

Ai, Bi to Ci = 40 ns
Ci to Si = $\underline{30 \text{ ns}}$
Add time   70 ns

Add time for a 32-bit adder:

Ai, Bi to Ci = 40 ns
Ci to Si = $\underline{30 \text{ ns}}$
Add time   70 ns

As can be seen, add time is independent of the number of bits added when using carry lookahead. This provides a significant improvement in add time, especially as the number of bits in the adder increases.

Unfortunately, providing full-carry lookahead for a large number of bits is quite complicated. For example, evaluating the expression for the most significant carry bit of an N-bit adder requires an $N + 1$ input OR gate and N AND gates ranging from 2 inputs to $N + 1$ inputs. For a 32-bit adder, this would require a 33-input OR gate and 32 AND gates ranging from 2 inputs to 33 inputs for just the most significant carry bit alone!

Because of this complexity, full-carry lookahead is typically done only 4 to 8 bits at a time. For example, consider a 32-bit adder comprised of four 8-bit adders, each with full-carry lookahead. In this arrangement, the carry must ripple between the four 8-bit adders, but the ripple effect is much less dramatic than full 32-bit ripple. (See Review Question 6.5.) This adder is shown in Figure 6.5.

6.3 Other Important Combinational Networks

There are several other combinational networks commonly found in digital circuits. These networks are easily designed using the same techniques we have covered so far. A few are mentioned here, their designs left to the reader as exercises.

6.3.1 Code Converters

Digital circuits generally represent "human" information via some sort of code. For example, BCD is simply a binary code representing the decimal values 0 through 9. It is often necessary in digital circuits to convert from one code to another. For

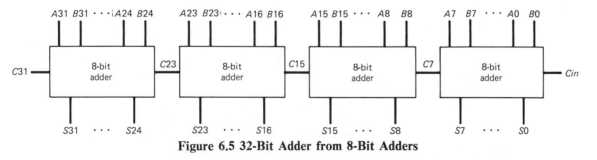

Figure 6.5 32-Bit Adder from 8-Bit Adders

example, consider designing a circuit to convert from BCD to 2421 code. BCD is sometimes referred to as 8421 code since the bit positions are weighted by 8, 4, 2, and 1. Likewise, 2421 code indicates bit-position weights of 2, 4, 2, and 1. From these definitions, a truth table for this conversion can be defined as shown in Table 6.3. This converter is easily built using reduction and implementation techniques covered so far (see Review Question 6.3). A 2421-to-BCD converter or any other converter can be designed in the same basic manner.

Complex code conversions involving more bits are usually implemented using PLAs (programmable logic arrays) and ROMs (read-only memories).

6.3.2 Comparators

It is often necessary to compare two binary values in digital circuits. Given two binary values, a *comparator* provides one or more outputs indicating equality, less than, greater than, and so on.

When designing a comparator, it is generally not necessary to derive expressions for each output as a function solely of the inputs. For example, if $A > B$ and $A < B$ are expressed as some function of the inputs, all other relations can be expressed as a function of these outputs. For example, A equals B only if A is not greater than B and B is not greater than A.

$$A = B \quad \Leftrightarrow \quad \overline{(A > B)} \cdot \overline{(B > A)}$$

Review Question 6.4 addresses the design of a comparator.

6.3.3 Display Decoder-Drivers

Digital circuits often "communicate" with the outside world via seven-segment LED displays such as those found in your calculator. These displays consist of seven individual LEDs, the various digits formed by lighting different LED segments. A display decoder-driver is used to properly light these segments based on an input value. For example, a BCD display decoder-driver has four inputs—the 4 bits of a BCD digit —and seven outputs—one for each LED segment. Based on the input BCD digit, the seven-segment outputs are properly set to form the digit on the display. (See Review Question 6.2.)

There are two basic configurations for seven-segment displays, *common anode* and *common cathode*. In a common-anode display, the anode or positive side of each LED is common, as shown in Figure 6.6. With these displays, an individual LED is lit by connecting the proper LED to ground. Thus a decoder-driver running a common-anode display sets its output low to turn on a segment. In a common-cathode

Table 6.3 BCD to 2421 Truth Table

BCD Input				2421 Output			
A	B	C	D	F1	F2	F3	F4
0	0	0	0	0	0	0	0
0	0	0	1	0	0	0	1
0	0	1	0	0	0	1	0
0	0	1	1	0	0	1	1
0	1	0	0	0	1	0	0
0	1	0	1	1	0	1	1
0	1	1	0	1	1	0	0
0	1	1	1	1	1	0	1
1	0	0	0	1	1	1	0
1	0	0	1	1	1	1	1

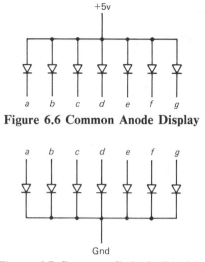

Figure 6.6 Common Anode Display

Figure 6.7 Common Cathode Display

display, the cathode or negative side of each LED is common, as shown in Figure 6.7. To light an individual LED on these displays, the proper LED is connected to a positive voltage. A decoder-driver running a common-cathode display sets its output high to turn on a segment.

The output of a TTL gate can supply more current than an LED can handle. To limit this current, a resistor is often placed between ground and each LED for common-anode displays, and between the positive supply and each LED for common-cathode displays.

Lab Exercise

Objective

This lab provides the student with an opportunity to use commercially available adders and display drivers. The final step results in a simple adder with binary inputs and a decimal readout.

Procedure   *p 96*

1. Design a circuit to add two 2-bit numbers using a 7483 4-bit adder. Use two switches on the logic designer to input one value and two more switches to input the other value. The 7483 adds two 4-bit values; what must be done with the unused inputs? What should be done with the carry input to the 4-bit adder? How many outputs from the adder must be connected to the LEDs on the logic designer to see all possible outputs?
   (a) Produce a logic diagram for this circuit. Show the adder as a single functional unit on your logic diagram. It has nine inputs—two 4-bit values and a carry input—and five outputs—the four sum bits and the carry out.
   (b) Verify circuit operation by checking all input combinations.   *p 93*
2. Design a circuit to display the decimal equivalent of a 4-bit BCD value, input via switches on the logic designer, on a seven-segment display. Use a 7448 BCD-to-seven-segment decoder to do this. The 7448 runs a common-cathode display and has built-in current-limiting resistors.
   (a) Produce a logic diagram for this circuit. Show the 7448 as a single functional unit on your diagram.
   (b) Verify circuit operation by trying all valid input combinations (0–9). What patterns are produced for inputs 10–15?
3. Using the circuits of (1) and (2), design a circuit to input two 2-bit numbers, add them, and display the result on the seven-segment display.

(a) Produce a logic diagram for this circuit.

(b) Verify circuit operation by trying all input combinations.

Have the lab instructor check your work after each circuit is working.

Review Questions

6.1 A 2-bit adder can be built from a half-adder and a full-adder as shown below:

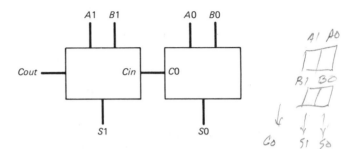

Design another 2-bit adder using the standard truth table approach. This truth table has four inputs: the 2 bits of A and the 2 bits of B. There are three outputs: the carry out *(Cout)* and the 2 bits of the sum. For example, shown here is minterm 6 of the truth table. The left 2 bits represent A, and the right 2 bits represent B. The outputs represent the sum of A and B and the resulting carry ($1 + 2 = 3$; i.e., 3 with a carry of 0).

$A2$	$A1$	$B2$	$B1$	$Cout$	$S2$	$S1$
0	1	1	0	0	1	1

(a) Produce the entire truth table.

(b) Produce the simplest equation for each output using K-maps for reduction. (Do not draw logic diagrams).

(c) How does this adder function differently from the adder illustrated here with respect to carry ripple?

6.2 Display decoder-drivers are available to properly light a seven-segment LED display based on a 4-bit BCD-digit input. A circuit is needed to display the hex representation of the remaining digits $(A–F)$ on the seven-segment LED. Let these digits be formed as:

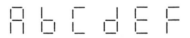

where segments are labeled:

(a) Produce a truth table illustrating the behavior of this circuit. This truth table will have four inputs—the 4 bits of the value to be displayed—and seven outputs—one for each segment of the LED. An output of 1 corresponds to a lit segment; an output of 0 corresponds to an unlit segment. For example, for an input of 10 (A), segments a, b, c, e, f, and g must be lit. Therefore, each of these outputs will be 1 for minterm 10. Since we are only interested in the values $A–F$, minterms 0–9 will all be don't cares.

(b) Produce the simplest equation for each output using K-maps to perform the reduction. (Do not draw logic diagrams.)

6.3 Referring to the truth table in Table 6.3, complete the design of a BCD-to-2421 code converter.
   (a) Produce the simplest equation for each output using K-maps to perform the reduction.
   (b) Produce logic diagrams using only NANDs and inverters to implement each equation.

6.4 A comparator circuit is needed to compare two 2-bit values, A and B. The outputs required are $A = B$, $A \neq B$, $A < B$, $A \leq B$, $A > B$, and $A \geq B$. These outputs should be 1 if the condition is true and 0 if the condition is false.
   (a) Illustrate how all remaining outputs can be derived from just two outputs: $A > B$ and $A < B$.
   (b) Produce a truth table for the comparator. This truth table will have four inputs, the 2 bits of A and the 2 bits of B, and two outputs, one for $A > B$ and one for $A < B$.
   (c) Using K-maps, produce the simplest equation for each of the two outputs, $A > B$ and $A < B$.
   (d) Produce a logic diagram using NANDs and inverters for this comparator showing implementation of the expressions from (c) and derivation of the remaining outputs from the outputs of (c).

6.5 Refer to the 32-bit adder formed from the 8-bit adders shown in Figure 6.5. Internally, each adder has full-carry lookahead producing a carry out 40 ns after all inputs are valid. Calculate the add time for the 32-bit adder.

6.6 Illustrate the way to add the following multibit binary numbers using a minimum of half- and full-adders:

$$
\begin{array}{cccc}
x8 & x4 & x2 & x1 \\
   &   & y2 & y1 \\
\hline
z8 & z4 & z2 & z1
\end{array}
$$

6.7 Design a decoder that will convert from 3-bit binary code to TEK code as shown. Show K-maps used and final equations.

Binary			TEK		
0	0	0	0	0	0
0	0	1	0	1	0
0	1	0	0	1	1
0	1	1	0	0	1
1	0	0	1	0	1
1	0	1	1	1	1
1	1	0	1	1	0
1	1	1	1	0	0

6.8 Implement the accompanying boolean functions using three half-adder circuits.
   (a) $F = A \oplus B \oplus C$ ($\oplus$ = exclusive or)
   (b) $W = \overline{A}BC + A\overline{B}C$
   (c) $F = AB\overline{C} + (\overline{A} + \overline{B})C$
   (d) $F = ABC$

6.9 Implement a full-adder circuit using two 4-to-1 muxes.

6.10 Design a combinational circuit whose input is a 4-bit number and whose output is the 2s complement of the input number.

6.11 Design a combinational circuit that multiplies an input BCD digit by 5 and expresses the answer as a two-digit BCD value. Show that the outputs can be obtained from the input lines without using any logic gates.

6.12 Design a combinational circuit that detects an error in the representation of a decimal digit in BCD. If the 4-bit input value is not a valid BCD digit, the output should be 1 to indicate an error, and 0 otherwise.

chapter 7

Latches and Flip-Flops

Objectives

1. An in-depth study of the operation of gated, master-slave, and edge-triggered latches and flip-flops.
2. An introduction to commercially available flip-flops.

7.1 Sequential Circuits

Circuits we have dealt with so far have all been *combinational* circuits. The output of a combinational circuit is a function solely of the inputs. The output of a *sequential* circuit is a function not only of the inputs but also of the current output; that is, there is feedback. Sequential circuits form the basis of registers, memories, and state machines, which in turn are vital functional units in digital design.

7.2 The *S-R* Latch

The most basic sequential unit is the *S-R* latch. From this basic circuit flip-flops are constructed, and from flip-flops, the registers, memories, and state machines just mentioned can be made. The basic *S-R* latch has two inputs, *S* and *R,* and two outputs, Q and $\overline{Q}$. Q and $\overline{Q}$ are always opposite; thus if Q is 1, $\overline{Q}$ is 0, and if Q is 0, $\overline{Q}$ is 1. The *S* input is used to (*S*)et Q to 1. The *R* input is used to (*R*)eset Q to 0.

Shown in Figures 7.1 and 7.2 are two implementations of the *S-R* latch using NOR gates and NAND gates respectively. As can be seen, the NOR implementation has active-high inputs. This indicates that the set or reset operation is performed by temporarily raising the set or reset line to 1. The NAND implementation has active-low inputs. This indicates the set or reset operation is performed by temporarily lowering the set or reset line to 0. Because of the feedback to the inputs of the OR forms, the output will remain set or reset even after the set or reset signal is gone—thus we have the memory capability of the latch. The reader is encouraged to verify this by following input sequences such as those illustrated in Figures 7.3 and 7.4 (pages 55 and 56) and determining the outputs of each gate in the latch.

Describing the behavior of sequential circuits is difficult, as the preceding paragraph of verbiage illustrates. Several methods exist for describing the nature of a sequential circuit, including state diagrams, state tables, transition tables, characteristic tables, and timing diagrams. In this chapter we introduce two methods: simple timing diagrams and reduced characteristic tables.

7.2.1 Timing Diagrams

The timing diagram provides the most exact depiction of circuit operation. Shown in Figures 7.3 and 7.4 are timing diagrams illustrating the behavior of the NOR and NAND implementations (respectively) of the *S-R* latch. The vertical axis is voltage,

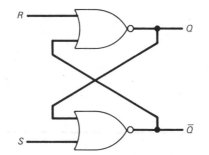

Figure 7.1 *S-R* Latch with NOR Gates

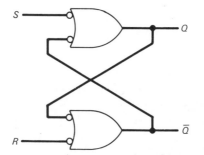

Figure 7.2 *S-R* Latch with NAND Gates

indicating the state of a line to be 0 or 1. The horizontal axis is time, illustrating the sequence of the voltage changes.

In Figure 7.3 the behavior of the NOR implementation is illustrated. Scanning from left to right, we see an initial state with Q, S, and R all 0. ($\overline{Q}$ will of course always be the opposite of Q.) The S(et) line is then raised to 1; this sets Q to 1, where it remains even after S returns to 0. Finally R(eset) is raised to 1; this resets Q to 0, where it remains even after R returns to 0. Figure 7.4 illustrates the behavior of the NAND implementation.

The timing diagrams shown in Figures 7.3 and 7.4 are simplified versions of the complex timing diagrams shown on manufacturers' data sheets. The preceding timing diagrams simply illustrate the function of the circuit, not the detailed timing of the electronics. For example, the state transitions shown actually take time; that is, the transitions are not vertical. In addition, it takes time from the transition of the set or reset line until the outputs actually change state. But since these times are so small, and since we are primarily interested in the *function* of the unit, not the electronic timing characteristics, we will deal with these simplified timing diagrams.

7.2.2 Characteristic Tables

While the timing diagram exactly illustrates the behavior of a circuit, it is fairly complex to read and not very compact. So far, we have used truth tables to illustrate

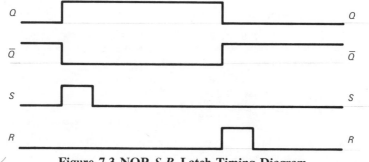

Figure 7.3 NOR *S-R* Latch Timing Diagram

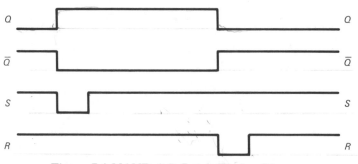

Figure 7.4 NAND *S-R* Latch Timing Diagram

the behavior of combinational circuits. A slightly modified version of a truth table, called a *reduced characteristic table,* can be used to illustrate the behavior of sequential circuits.

Shown in Tables 7.1 and 7.2 are characteristic tables for the NOR and NAND implementations of the *S-R* latch. The table answers the question, "What will the new state of the output be if the inputs are set to these values?" The current state of the output is referred to as Q. The next state of the output, that is, after the transition is made by changing the inputs, is referred to as $Q+$. For example, from the first line of the table we can see that if both inputs are set to 0, then the new output $Q+$ will be Q; that is, the new output will remain the same as it was before. From line 3 we can see that if S is set to 1, the new state of the output $Q+$ will be 1. If S is then returned to 0, line 1 of the table shows the new output $Q+$ will be the same as before. In this case, since Q was 1, the new state $Q+$ will also be 1.

Note the invalid condition in the NOR implementation when both S and R are 1 and in the NAND implementation when S and R are both 0. Looking at the NOR implementation of the *S-R* latch in Figure 7.1, we can see why this condition is invalid. Under normal conditions, the outputs of the two NOR gates are always opposite (Q and $\overline{Q}$). If S and R were both set to 1, then the outputs of both gates would be 0 —no longer opposite. In addition, when S and R return to 0, the state the circuit would settle into is not definite.

7.3 Synchronous Latches (Flip-Flops)

The output of the *S-R* latch discussed so far will change at the same instant the inputs dictate a change. This is referred to as *asynchronous* operation. But in most digital circuits, the timing of events is critical. Typically a clock is used to synchronize

Table 7.1 NOR *S-R* Latch Characteristic Table

S	R	Q+	
0	0	Q	(No change)
0	1	0	(Reset output)
1	0	1	(Set output)
1	1	—	(Invalid condition)

Table 7.2 NAND *S-R* Latch Characteristic Table

S	R	Q+	
0	0	—	(Invalid condition)
0	1	1	(Set output)
1	0	0	(Reset output)
1	1	Q	(No change)

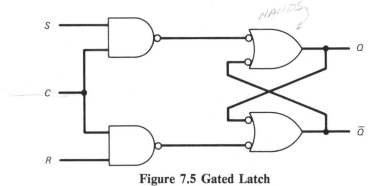

Figure 7.5 Gated Latch

the operation of the various functional units in the circuit. To perform in this environment, the latch must only change state when a clock signal indicates that it is time.
This is referred to as *synchronous* operation. Latches modified to operate synchronously
are called flip-flops. How can synchronous operation be achieved?

7.3.1 Gated Latches

To achieve synchronous operation, the latch should change state only on the
proper clock signal. For example, assume that the latch should change state only when
the clock signal goes high. The simplest approach to achieve this is the addition of a
gate at the inputs that permits the latch to "look" at the inputs only when the clock
is high. This is referred to as a *gated latch* and is illustrated in Figure 7.5.

The left two NANDs form the gate; the right two NANDs form the latch, as
shown earlier. While the clock input (C) is low, the outputs of the gate are always 1
regardless of the inputs S and R. When the clock (C) goes high, then the S and
R inputs are passed to the latch and the output changes. The behavior of this circuit
is shown in Table 7.3.

There is a slight problem with this solution. During the time the clock is high,
the gated latch performs identically to the regular asynchronous latch. Thus, if the
inputs changed multiple times while the clock was high, the state of the latch could
also change multiple times. This problem is illustrated in the timing diagrams shown
in Figure 7.6. The first clock cycle is a "normal" one in which the set line goes high
and Q follows as soon as the clock goes high. While the clock is high, S returns to 0,
and no other change in inputs occurs. The second clock cycle illustrates the problem.
First the reset line goes high, and Q is set to 0 as soon as the clock goes high. But while
the clock is still high, the set line goes high, and Q is set back to 1 during the same
clock cycle.

7.3.2 Edge-triggered Flip-Flops

The most common solution to this problem is the edge-triggered flip-flop. In a
rising edge-triggered flip-flop, the inputs are looked at only on the rising edge of the

Table 7.3 Gated Latch Characteristic Table

C	S	R	$Q+$	
0	0	0	Q	(Output remains the
0	1	0	Q	same while clock
0	0	1	Q	is low
0	1	1	Q	
1	0	0	Q	(No change)
1	0	1	0	(Reset output)
1	1	0	1	(Set output)
1	1	1	—	(Invalid condition)

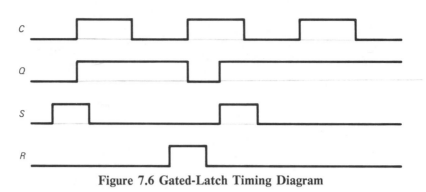

Figure 7.6 Gated-Latch Timing Diagram

clock. On a falling edge-triggered flip-flop, the inputs are looked at only on the falling edge of the clock. For example, in a rising edge-triggered flip-flop, the inputs are latched to the outputs on the rising edge of the clock and ignored until the next rising edge.

The edge-triggered flip-flop can easily be made from the gated latch previously discussed. A pulse-narrowing circuit is added before the gate control input so that the gate is active for only an instant on the rising or falling edge of the clock. This way, there is no time for the inputs to multiply change states while the gate is enabled. This implementation is illustrated in Figure 7.7.

Note the addition of two more inputs, $\overline{\text{Preset}}$ and $\overline{\text{Clear}}$. Since these inputs run directly to the latch, they can immediately set or reset the output independent of the clock. They are referred to as *asynchronous inputs* and are useful for initialization.

There are several ways to implement the pulse-narrowing circuit just shown. One simple implementation is given in Figure 7.8. This method uses the fact that a gate actually takes time to switch states. Assume an initial state with the clock input low. Because of the input inverter, inputs A and B on the NAND gate will be opposite, the output of the NAND gate will be 1, and the final output will be 0. When the clock goes high, input B on the NAND gate immediately goes high, but input A also remains high until the inverter switches from high to low. This takes approximately 10 ns with standard TTL. Thus for 10 ns, both inputs to the NAND gate are high, the output of the NAND gate is low, and the output of the circuit is high. After 10 ns, the input inverter produces a low output, which in turn changes the output of the NAND gate to 1, and the final output returns to 0. (see Review Question 7.5).

7.3.3 Master-Slave Flip-Flops

Another technique for eliminating multiple-state transitions during a single clock cycle is the use of a master-slave arrangement. A master-slave flip-flop is formed from

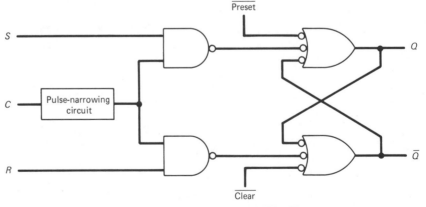

Figure 7.7 Edge-Triggered Flip-Flop

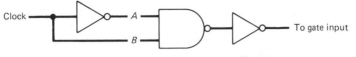

Figure 7.8 Pulse-Narrowing Circuit

two gated latches, as shown in Figure 7.9. The left or master latch forms the inputs to the flip-flop, and the right or slave latch forms the outputs of the flip-flop. The master latch looks at the inputs while the clock is high. When the clock returns low, the slave latch is enabled, using the outputs of the master latch as its inputs. Thus the inputs are "read" while the clock is high and transferred to the outputs when the clock returns low.

7.4 TTL Flip-Flops

There is no need to construct flip-flops from individual gates since many variations exist on standard TTL chips. There are typically two to eight flip-flops in a single package; the most popular are discussed in the following sections.

7.4.1 *J-K* Flip-Flops

The *J-K* flip-flop is simply an *S-R* flip-flop that has been modified so that both inputs can be active at the same time. Where in the *S-R* flip-flop this condition was considered invalid, in the *J-K* flip-flop this condition toggles the output on successive clock cycles. This behavior and the standard representation of the *J-K* flip-flop is illustrated in Figure 7.10.

The *J-K* flip-flop is available in several TTL chips with various options. For

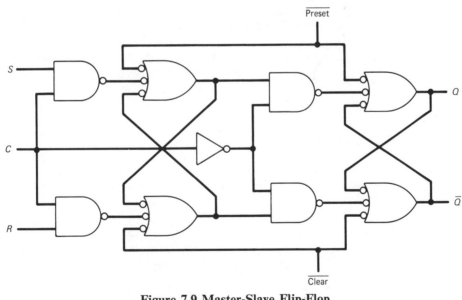

Figure 7.9 Master-Slave Flip-Flop

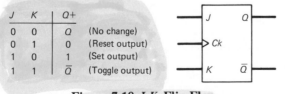

J	K	Q+	
0	0	Q	(No change)
0	1	0	(Reset output)
1	0	1	(Set output)
1	1	$\overline{Q}$	(Toggle output)

Figure 7.10 *J-K* Flip-Flop

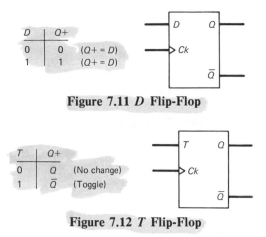

D	Q+	
0	0	(Q+ = D)
1	1	(Q+ = D)

Figure 7.11 *D* **Flip-Flop**

T	Q+	
0	Q	(No change)
1	Q̄	(Toggle)

Figure 7.12 *T* **Flip-Flop**

example, the 7473 contains two *J-K* flip-flops with an asynchronous Clear input, as mentioned previously. The 7476 contains two *J-K* flip-flops with both Preset and Clear.

7.4.2 *D* Flip-Flops

The *D* or Data flip-flop has just one control input, *D*. The output of the flip-flop is set to whatever the *D* input is on each clock cycle. This behavior and the standard representation of the *D* flip-flop is shown in Figure 7.11.

As with *J-K* flip-flops, *D* flip-flops are also available on standard TTL chips. For example, the 7474 contains two *D* flip-flops with Preset and Clear.

7.4.3 *T* Flip-Flops

The *T* or Toggle flip-flop has just one control input, *T*. If this input is 1, the output toggles on successive clock cycles. If the *T* input is 0, the output remains unchanged on successive clock cycles. This behavior and the standard representation of the *T* flip-flop are shown in Figure 7.12.

The *T* flip-flop is easily made from the *J-K* flip-flop by tying the *J-K* inputs together. Thus there are no commercially available *T* flip-flops.

Lab Exercise

Objective

This lab exposes the student to the operation of simple sequential devices. The invalid condition in an *S-R* latch, the rising and falling edge-triggered nature of various flip-flops, and the effects of switch bounce are observed.

Procedure

1. Wire the basic NOR implementation of the *S-R* latch as shown in Figure 7.1.
   (a) Verify proper operation of the latch by alternately setting and resetting the output using the *S* and *R* inputs.
   (b) Set both *S* and *R* to 1; what are the outputs?
   (c) Return *S* and *R* to 0 "simultaneously." Repeat (b) and (c) several times. Do the outputs return to the same state each time *S* and *R* are returned to 0?
2. Connect the *J-K* inputs of a 7476 *J-K* flip-flop to the input switches on the logic designer. Connect the clock input of the flip-flop to one of the pulser switches and the outputs of the flip-flop to the LEDs.
   (a) Verify proper operation by trying each *J-K* combination and pressing the pulser to clock the flip-flop. Note that changing the *J-K* inputs does not affect the state of the flip-flop until the flip-flop is clocked.

(b) When does the flip-flop change state—on the rising or falling edge of the clock?

(c) Connect the clock input of the *J-K* flip-flop to an input switch on the logic designer. Observe the problem described in Review Question 7.3 by setting the *J* and *K* inputs to 1 and clocking the flip-flop using the input switch.

(d) Produce a timing diagram showing the clock input, the *J-K* inputs, and the outputs as the sequence in (a) is performed.

3. Connect the *D* input of a 7474 *D* flip-flop to an input switch on the logic designer. Connect the clock input to one of the pulser switches and the outputs to the LEDs.

(a) Verify proper operation by setting the input switch to 1 and 0 and pressing the pulser to clock the flip-flop.

(b) When are the outputs set—on the rising or falling edge of the clock?

(c) Produce a timing diagram showing the clock input, the *D* input, and the outputs as the sequence in (a) is performed.

Have the lab instructor check your work after each circuit is completed.

Review Questions

7.1 Illustrate how a *D* flip-flop can be made from a *J-K* flip-flop with external gating.

7.2 Illustrate how a *T* flip-flop can be made from a *D* flip-flop with external gating.

7.3 A "clock" pulse for a flip-flop could be generated by manually flipping a switch from 0 to 1 and then back to 0. But switches such as those on the logic designer "bounce" when switched from one state to another. For example, when switching to 1, the output actually bounces several times between 1 and 0 before finally settling to 1. What problems could switch bounce cause in applications like this one?

7.4 In Figure 7.7, two new inputs were introduced, the Preset and Clear inputs. What must be done to set the output to 1? To 0?

7.5 Refer to the pulse-narrowing circuit shown in Figure 7.8. Produce a timing diagram showing the input clock, *A, B,* the output of the NAND gate, and the final output of the circuit for one input-clock cycle. Show and label the actual delays between the signals, assuming each gate has a switching delay of 10 ns.

7.6 Design a pulse-narrowing circuit for a falling edge-triggered device. On the falling edge of the input clock, a short pulse should be output.

7.7 How would the circuit of Figure 7.2 operate if the *S* input of the NAND gate became disconnected?

7.8 How would the circuit of Figure 7.1 operate if the *S* input of the NOR gate became disconnected?

7.9 For the timing diagram in the accompanying figure, sketch the output if the input is:

(a) the *D* input to a 7474

(b) The *D* input to a gated *D* latch (a *D* flip-flop with a gate enable similar to Figure 7.6 rather than a clock)

(c) The *J* input to a master-slave *J-K FF* with the *K* input tied HIGH

(d) The *J* input to a rising edge-triggered *FF* with the *K* input tied HIGH

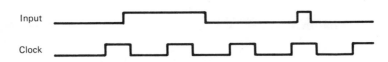

chapter

8

Counters

Objectives

1. Applications of counters in digital circuits.
2. A detailed study of ripple-counter operation and limitations imposed by switching delays and glitches caused by the asynchronous nature of the ripple counter.

8.1 Counters Defined

In general terms, a counter is a sequential circuit that goes through a set of given states on successive clock cycles. Usually they are sequential binary states (e.g., 0 through 7, 0 through 15, and so on); hence the term *counter.* A state is simply the binary value represented by the outputs of the counter. For example, a counter to sequence from 0 through 15 requires four outputs to represent the four bit positions.

Counters are useful in many applications, including timing, multiplexing, and, of course, counting. In this chapter we study the design and application of counters, including commercially available counters.

8.2 Asynchronous Ripple Counters

If a *J-K* flip-flop is used as a *T* flip-flop (by tying *J* and *K* high), the output changes state on each clock cycle. Therefore, after two clock cycles the output of the flip-flop completes one cycle. If the output of this flip-flop were then used as the clock input to a second *T* flip-flop, the output of the second flip-flop would cycle at half the rate of the first flip-flop. Each additional flip-flop added in this manner would cycle at half the rate of the preceding flip-flop. This behavior for two flip-flops is illustrated in the timing diagram shown in Figure 8.1.

As can be seen, by considering the output of the first flip-flop (Qa) as the 1s digit

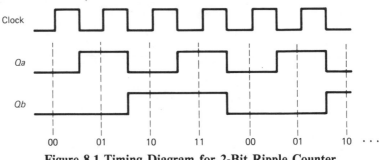

Figure 8.1 Timing Diagram for 2-Bit Ripple Counter

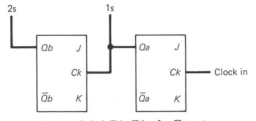

Figure 8.2 2-Bit Ripple Counter

and the output of the second flip-flop (Qb) as the 2s digit, we have a simple binary counter sequencing from 0 through 3 and then repeating. A circuit to implement this counter is shown in Figure 8.2. The absence of connections to the J and K inputs indicates they are tied high. A counter to sequence from 0 through 7 can be made by adding an additional flip-flop (3 in total); a counter to sequence from 0 thru 15 by adding two additional flip-flops (4 in total), and so on.

Note from the timing diagram in Figure 8.1 that state transitions occur on the falling edge of the clock. Recall from the lab in Chapter 7 that the 7476 J-K flip-flop also changes state on the falling edge of the clock. These flip-flops are well suited for making counters.

Also note that the timing diagram shows state transitions occurring exactly on the falling edge of the clock. As mentioned in Chapter 7, this isn't actually true; it takes time from the falling edge of the clock until the output of the flip-flop changes state. For the 7476 flip-flop this delay is approximately 40 ns. For the 2-bit ripple counter shown in Figure 8.2, the 1s digit (Qa) changes 40 ns after the falling edge of the clock. Since Qa is the clock input for the second flip-flop, the 2s digit (Qb) changes 40 ns after the falling edge of Qa and 80 ns after the original clock edge. With each additional flip-flop added, a 40 ns switching delay is incurred as the clock signal "ripples" through the string of flip-flops. Since each flip-flop does not change state on a common clock signal, it is not considered a synchronous device. Put these observations together and the origin of the name *asynchronous ripple counter* is clear. Asynchronous ripple counters are often referred to as *ripple counters*, for short.

8.2.1 Timing Considerations

The maximum speed at which a ripple counter can count depends on the switching delay of the flip-flops used and the number of flip-flops in the counter. For example, consider a 4-bit ripple counter using the 7476 J-K flip-flops previously mentioned. The longest delay occurs when all four flip-flops from the least significant (Qa) to the most significant (Qd) must change state. This occurs when switching from state 7 (0111) to state 8 (1000) or when switching from state 15 (1111) to state 0 (0000). Since each flip-flop introduces a 40 ns delay, the total delay to switch from one state to another and the resulting maximum counting rate are:

maximum switching delay $= 4 \times 40$ ns $= 160$ ns
maximum counting rate $= 1/160$ ns $= 2.25$ MHz

This value for maximum counting rate assumes that the application requires that each state exist for some short time before the next clock pulse. A practical limit on the counting rate would of course be lower since at this rate states 0 and 8 would exist for such a short time (see Review Question 8.3). The timing diagram in Figure 8.3 shows the switching delay incurred when switching from state 15 to state 0.

The output states of a counter are often used as control inputs to multiplexors, decoders, memories, and so on. The state might represent a memory location to read

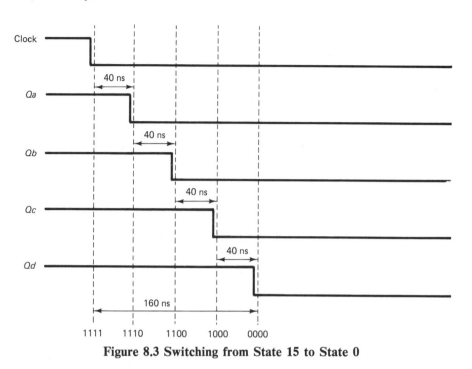

Figure 8.3 Switching from State 15 to State 0

or a decoder output to set to 0. But with a ripple counter, the transition from one state to another is not instantaneous. Figure 8.3 shows the sequence of flip-flop transitions as a ripple counter switches from state 15 (1111) to state 0 (0000). At the falling edge of the clock, the output state is 15 (1111). Forty ns later, Qa falls to 0, changing the output to 14 (1110). Forty ns after Qa, Qb falls to 0, leaving the output at 12 (1100). Qc falls to 0 40 ns after Qb, leaving the output at 8 (1000), and finally Qd falls to 0 40 ns after Qc, 160 ns after the original clock edge, with the output now at 0. So as the counter switches from state 15 to state 0, states 14, 12, and 8 exist for 40 ns each. These intermediate states are often referred to as *glitches* or *decoding spikes*. Switching from state 15 to state 0 is not the only time that glitches occur. Any time more than one flip-flop must change state, intermediate states will exist (see Review Question 8.4).

To illustrate the effect of glitches, consider a counter sequencing the select lines of a decoder from 0 through 15. The outputs of the decoder might be turning on one of 16 different seven-segment displays. But as previously described, when switching from state 15 to state 0, states 14, 12, and 8 also exist for 40 ns each. Thus in this application, displays 14, 12, and 8 will each turn on for 40 ns when switching from display 15 to display 0. Of course a 40 ns flicker of a display is not evident to the human eye, but if these decoder outputs are being read by another digital device, 40 ns can be a long time!

In such an application where the output from the decoder, or possibly a multiplexor, must be read by other digital devices, timing is important. Obviously the desired data is not valid until all intermediate states have passed and the counter has settled into the final state. It is important therefore that the "reading" device allow enough time for the counter to settle into the final state.

8.3 Generating Clock Signals with Ripple Counters

A ripple counter can be used to divide the rate of one clock signal to generate another clock signal of the desired frequency. For example, assume a digital circuit has a clock running at 10 KHz. In addition to the 10 KHz signal, the circuit also requires a 5 KHz clock. Instead of having separate clock circuits for each of these frequencies, we can obtain the 5 KHz clock by dividing the 10 KHz signal by 2.

Recall from Section 8.2 that the output of a T flip-flop cycles at half the rate of the input clock. If a 10 KHz signal clocks the flip-flop, the output cycles at one-half that rate, or 5 KHz. Thus a 1-bit ripple counter performs a divide-by-2 operation. Also recall from Section 8.2 that if a second T flip-flop is added, it cycles at half the rate of the first flip-flop, one-fourth the rate of the original clock. Thus a 2-bit ripple counter performs a divide-by-4 operation. The "divide-by-n" operation that a counter performs equals the number of states in the counter's sequence. For example, a 1-bit ripple counter has two states, 0 and 1, and performs a divide-by-2 operation. A 2-bit ripple counter has four states, 0 through 3, and performs a divide-by-4 operation. Similarly, a 3-bit counter has eight states, 0 through 7, and performs a divide-by-8 operation.

8.3.1 Modifying the Count Sequence

The number of states in a ripple counter's sequence has always been a power of 2 so far. How could a clock signal be divided by some value not a power of 2—5, for example? Obviously to divide by 5, the counter must have five states. To do this, we use a regular ripple counter, but with additional gating, we force it to reset to 0 after five states. In this case, the counter will count 0, 1, 2, 3, 4 and then repeat. Somehow, instead of going from state 4 to state 5, the ripple counter must be forced to go from state 4 to state 0.

Recall from the lab in Chapter 7 the presence of asynchronous-clear inputs on the 7476 J-K flip-flop. Using these inputs, the ripple counter can be reset to 0 when needed. In this example, it must reset to 0 when the ripple counter attempts to switch to state 5. A divide-by-5 counter is shown in Figure 8.4. As can be seen, state 5 is detected by a NAND gate when Qc and Qa are both 1. At this instant the NAND gate turns on and resets the counter to state 0. This operation results in the complex timing diagram shown in Figure 8.5. Forty ns after the falling edge of the clock, Qa goes to 1, leaving the counter in state 5. At this instant, both inputs to the NAND gate are 1, and 10 ns later (the switching delay of the NAND gate), the output of the NAND gate goes low to reset the flip-flops. The clear operation on a 7476 J-K flip-flop takes a maximum of 40 ns. Thus the output is finally reset to 0 90 ns ($40 + 10 + 40$) after the falling edge of the clock. Note that state 5 does exist for 50 ns, but looking at Figure 8.5 we can see the intermediate state does not effect Qc—the divide-by-5 output.

8.3.2 Generating a Symmetric Waveform

The resulting square wave generated by the divide-by-5 counter previously described is not symmetric. It is 0 for four input-clock cycles (000 thru 011) and 1 for one input-clock cycle (100). Figure 8.6 shows the input clock and the divide-by-5

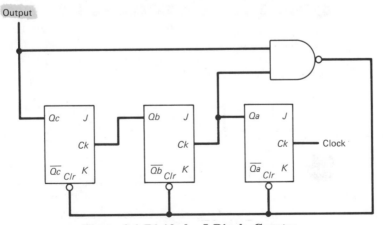

Figure 8.4 Divide-by-5 Ripple Counter

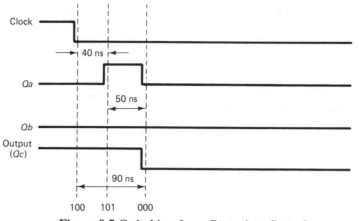

Figure 8.5 Switching from State 4 to State 0

output taken from Qc. The numbers across the bottom indicate the current state of the counter after each transition. Switching delays are not shown.

Some applications require a more symmetric waveform: one where the output is 1 the same amount of time it is 0. Obviously this is impossible when dividing the clock by an odd value, but a more symmetric waveform can often be obtained by taking the output from a different counter output. For example, the following shows the state sequence of the divide-by-5 counter previously described.

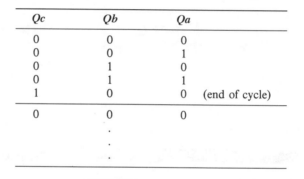

Qc	Qb	Qa	
0	0	0	
0	0	1	
0	1	0	
0	1	1	
1	0	0	(end of cycle)
0	0	0	
	.		
	.		
	.		

As can be seen, Qc cycles once during the sequence, but Qb also cycles only once during the sequence. Since Qb cycles only once, it too can be used as the output. In addition, Qb is more symmetric: it is 0 for three cycles (4, 0, and 1) and 1 for two cycles (2 and 3). Which counter outputs can be used as the clock output and which are most symmetric depend on the sequence length (see Review Question 8.5).

It is important that the counter output used as the clock output is not affected by glitches. For example, in the previous sequence, Qa appears to remain 0 between state 4 and state 0. But actually, as shown in Figure 8.5, Qa temporarily goes to 1 when switching from state 4 to state 0. If Qa were used as a clock output, this short 40 ns pulse could cause several problems.

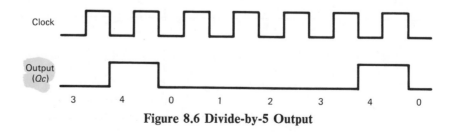

Figure 8.6 Divide-by-5 Output

If the desired sequence length is even, a completely symmetric waveform can be generated. A divide-by-2 operation always produces a symmetric waveform. By breaking an even divide operation into two divide operations where the last is a divide-by-2, a symmetric waveform will result. For example, a divide-by-10 operation can be accomplished by dividing by 5 and then dividing this signal by 2. The final divide-by-2 operation will produce a symmetric waveform. This technique requires no additional flip-flops since although one extra is required to do the divide-by-2 operation, one less is required to do the original divide operation. Figure 8.7 shows a symmetric divide-by-6 ripple counter. The right two flip-flops form the divide-by-3 counter. Note that the NAND gate resets the counter to 0 when switching from state 2 to state 3. The leftmost flip-flop is the divide-by-2 counter from which the output is taken.

8.4 Other Applications of Counters

Generating clock signals is certainly not the only application of counters. Counters are used in many applications, including multiplexing applications, pulse timing, pulse counting, frequency measurement, and analog-to-digital converters.

In multiplexing applications, the ripple counter is generally used to provide the select-line sequence for the multiplexor or decoder. The effect of glitches as covered in Section 8.2.1 must be carefully considered in multiplexing applications, especially when multiplexing is used to pass data to other digital devices.

A counter can easily be started and then stopped by two consecutive input pulses (see Review Question 8.6). By clocking the counter at a known rate between the pulses, the time between them can be determined. In these applications, the number of bits in the counter and the maximum counting rate are the primary design considerations.

A counter can also count pulses. In these applications, the pulses are used as the clock input for the counter. By counting over a given period of time, the average frequency of the pulses can be determined. Primary considerations in this application are again the number of bits in the counter and maximum counting rate.

An analog-to-digital converter converts an input voltage into a binary value representing that voltage. For example, an 8-bit A/D converter might be used to convert a voltage between 0 and 5 volts into a binary value between 0 and 255. *A*-to-*D* conversions can be done in many ways, and several of those methods use counters. For example, one simple method has the outputs of a counter feeding the inputs of a digital-to-analog converter. For each output of the counter, the D/A converter will output a different voltage. A device to compare voltages called a *comparator* is used to determine when the output of the D/A converter is greater than the input voltage.

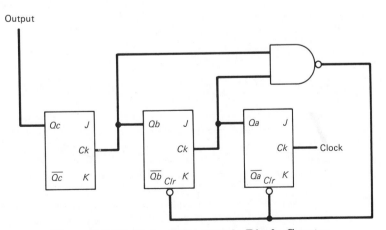

Figure 8.7 Divide-by-6 Symmetric Ripple Counter

When it is, the counter is stopped, and the value in the counter represents the binary value of the input voltage.

8.5 Commercially Available Ripple Counters

It is rarely necessary to build counters from individual flip-flops since several counters are available on chips. For example, the 7493 is a 4-bit ripple counter using J-K flip-flops. Three of the flip-flops are internally connected as a 3-bit ripple counter. A fourth flip-flop has a separate clock input, and its output is not connected to the other flip-flops. To use the chip as a 3-bit counter, we provide the clock signal to the clock input of the 3-bit counter. To use the chip as a 4-bit counter, we connect the output of the single flip-flop to the clock input of the 3-bit counter, and we provide the clock signal to the single flip-flop (see the data sheet for the 7493 in Appendix B). A common clear line runs to all flip-flops, but instead of there being simply an active-low clear such as that found on the 7476 flip-flop, the clear operation is gated by a two-input NAND gate. To clear the counter, both inputs to the internal NAND gate must be set to 1. Notice in Figure 8.4 that this is exactly how we modified the count sequence of a 3-bit ripple counter to obtain a divide-by-5 counter. The divide-by-5 counter can easily be implemented with just the 7493 chip, whereas without it, two 7476 chips and one 7400 chip are required.

Other counters available have different count sequences. The 7490 is a decade counter. It has ten states, 0 thru 9, and is extremely useful in decimal counting applications. Its count sequence has been shortened to 0 through 9 using a different method than discussed in Section 8.3.1 (see the data sheet for the 7490 in Appendix B). Like the 7493, the 7490 is split into a 3-bit counter and a single separate flip-flop. Using the 3-bit counter, the 7490 performs a divide-by-5 operation (five states). Using it as a 4-bit counter, the 7490 performs a divide-by-10 operation (ten states). The counter is reset in the same manner as the 7493 with an internal NAND gate. The 7490 can also be set to 9 through an internal NAND gate.

Many other counters are available; the 7492 is a divide-by-12 or divide-by-6 counter. The 74163 is a synchronous 4-bit binary counter. In a synchronous counter, all outputs change state at the same time. The design of synchronous counters and state sequencers that don't necessarily count sequentially will be covered in Chapter 9.

Lab Exercise

Objective

This lab exposes the student to the use of ripple counters. J-K flip-flops are used to make a simple ripple counter, and then a 7493 is used to make two versions of a divide-by-10 counter. To illustrate multiplexor and decoder applications, a circuit to count on a seven-segment LED and a *goofy-lights* display are made.

Procedure

1. Using 7476 J-K flip-flops, design and wire a simple 4-bit counter sequencing from 0 through 15. Hook the outputs of the counter to the LEDs on the logic designer. Produce a logic diagram for this counter.

2. A circuit is needed to divide a 10 Hz signal down to 1 Hz. Design and wire two circuits as in the accompanying description. Hook the clock input to a 10 Hz clock on the logic designer, and hook the output clock to an LED. Produce a logic diagram for both designs:
   (a) Using only a 7493 4-bit counter. Note the asymmetric output on the LED.
   (b) Using a 7493 4-bit counter and a 7476 J-K flip-flop to provide a symmetric square-wave output. Why can't the single flip-flop in the 7493 be used as the divide-by-2 counter?

3. Using a 7490 decade counter, design and wire a circuit to count from 0 through 9 and display the values on a seven-segment display. Produce a logic diagram for your circuit.
4. In the lab in Chapter 5 a circuit was made to light one of four LEDs based on a 2-bit input combination. Design a similar circuit using a 2-bit counter made from 7476 *J-K* flip-flops that automatically sequences through all four input combinations. At higher frequencies, the light will appear to be moving across the four LEDs. Produce a logic diagram for your circuit.

Have the lab instructor check your work after each circuit is working.

Review Questions

8.1 Design a 2-bit ripple counter using *D* flip-flops.
8.2 The *modulus* of a counter is the number of states before the counter repeats, that is, the number of states in the counter's sequence. What is the modulus of a 5-bit binary counter? How many flip-flop stages are required for building a modulo-256 counter?
8.3 Assuming a 4-bit binary ripple counter made from *J-K* flip-flops with timing characteristics as given in this chapter, how long does state 8 exist if clocking at 2.0 MHz? How long does state 0 exist at 2.0 MHz?
8.4 For a 3-bit binary ripple counter, which state transitions produce decoding spikes?
8.5 Design a divide-by-9 counter using four *J-K* flip-flops. The output of the counter should provide the most symmetric waveform possible.
8.6 Design a circuit to start and stop a counter on the rising edge of two consecutive input pulses. This circuit can be used to measure the time between the two pulses, as described in Section 8.4. To accomplish this, your circuit should turn the clock for the counter on and off as shown in the following illustration. On the rising edge of the first input pulse, the clock signal is passed to the counter. On the rising edge of the second input pulse, the clock signal to the counter is turned off. Ignore the problem of synchronizing with the input clock.

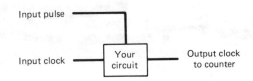

8.7 Design a divide-by-12 counter using four *J-K* flip-flops. What is the ratio of the time the output is high to total cycle time (duty cycle)?
8.8 Repeat Question 8.7 to obtain a 50 percent duty cycle.
8.9 What intermediate states exist when an 8-bit counter switches from state 127 to state 128?
8.10 Assuming the counter of Question 8.9 is made from *J-K* flip-flops with timing characteristics as given in this chapter, what is the delay from the clock edge until state 128 is valid?
8.11 Show how to connect a 7493 4-bit ripple counter to count the full sequence from 0 through 15.
8.12 Show how to connect a 7493 4-bit ripple counter to form a divide-by-12 counter. Are any external gates required?
8.13 Construct a 2-bit ripple counter using one *D* flip-flop and one *J-K* flip-flop.

chapter 9

State Sequencers and Controllers

Objectives

1. An introduction to the design of state sequencers used in controlling digital circuits.

2. The use of transition tables and state diagrams in designing and analyzing these circuits.

9.1 State Sequencers and Controllers Defined

In Chapter 8 we saw various ways counters can be used to control digital circuits. But often the control requirement extends beyond the simple ascending sequence of states that a counter provides. Sequential circuits that can generate any specific sequence of states are referred to as *state sequencers* or *controllers*.

Unlike ripple counters, state sequencers are generally synchronous; that is, all outputs change at the same time based on a common clock. Obviously a state sequencer that generates an ascending sequence of states is simply a synchronous counter. But with the ability to generate other state sequences, state sequencers can provide a more complex control sequence. The output of a state sequencer is often broken into various size fields, each field controlling an operation that must take place. In a simple arithmetic processor, for example, individual fields might control register loading, data routing via select lines on a multiplexor, function selection on an ALU, and so on.

To aid in the design of these elements, we introduce two new methods of illustrating sequential circuit behavior, the state diagram and the transition table.

9.1.1 State Diagrams

A state diagram provides a visual representation of sequential circuit behavior. A state diagram is a directed graph with the nodes indicating the possible output states and the arrows representing valid state transitions. Often the arrows are annotated with the input conditions that cause that particular transition path to be taken. Specific state diagrams will be illustrated in the following design examples.

9.1.2 Transition Tables

A transition table lists all possible state transitions and the inputs required to produce that change. Shown in Table 9.1 is the transition table for a *J-K* flip-flop. The second line, for example, shows that to change the output from 0 to 1, *J* must be 1 and *K* can be either 0 or 1 (don't care). More complex transition tables will be illustrated in the following design examples.

Table 9.1 *J-K* Flip-Flop Transition Table

Present State	Next State	Inputs	
Q	$Q+$	J	K
0	0	0	x
0	1	1	x
1	0	x	1
1	1	x	0

9.2 Synchronous Counters

As described in Chapter 8, ripple counters have many uses. In some applications, though, the glitches caused by the asynchronous nature of the counter can be a problem. In this section, the design of synchronous counters is covered. In a synchronous counter, all flip-flops change state off a common clock signal. Therefore, all flip-flops change state at the same time and thus eliminate glitches.

Assume a 3-bit binary counter sequencing from 0 through 7 is needed. To design a synchronous counter we start with one flip-flop for each bit, as in a ripple counter. But instead of obtaining the clock input from the previous flip-flop, we have all flip-flops run off a common clock and add simple logic to generate the proper sequence. In this application, three flip-flops are needed to represent states 0 through 7 (000–111).

The sequence of state transitions and the *J-K* inputs required to make those transitions are shown in Table 9.2.

The *J-K* values shown are those required to make each flip-flop perform the proper state transition. For example, the first state transition is from 000 to 001. In this case, flip-flop outputs Qc and Qb must remain at 0, and Qa must change from 0 to 1. Looking at the transition table for the *J-K* flip-flop in Table 9.2, we see that *J* must be 0 and *K* a don't care in order to maintain an output of 0 for Qc and Qb. For Qa to change from 0 to 1, *J* must be 1, and again *K* is a don't care.

For each present state there is a different set of *J-K* inputs required to properly perform the state transition to the next state. Thus the *J-K* inputs can be thought of as a function of the present state, that is, of Qc, Qb, and Qa. In this light, we derive the boolean expression for each *J* and *K* input as a function of Qc, Qb, and Qa. Shown in Figure 9.1 are the K-map reductions and final equations for each of the *J-K* inputs. Shown in Figure 9.2 is the logic diagram implementing this sequencer.

9.2.1 Commercially Available Synchronous Counters

Several variations of the synchronous counter are available on TTL chips. Many of these can be parallel loaded to start the counter at a specific state. Parallel and serial

Table 9.2 Transition Table for a 3-Bit Counter

Present State			Next State			Flip-Flop Inputs to Perform Change					
Qc	Qb	Qa	Qc	Qb	Qa	Jc	Kc	Jb	Kb	Ja	Ka
0	0	0	0	0	1	0	x	0	x	1	x
0	0	1	0	1	0	0	x	1	x	x	1
0	1	0	0	1	1	0	x	x	0	1	x
0	1	1	1	0	0	1	x	x	1	x	1
1	0	0	1	0	1	x	0	0	x	1	x
1	0	1	1	1	0	x	0	1	x	x	1
1	1	0	1	1	1	x	0	x	0	1	x
1	1	1	0	0	0	x	1	x	1	x	1

(handwritten margin note:)

J	K	Q+
0	0	Q
0	1	0
1	0	1
1	1	$\overline{Q}$

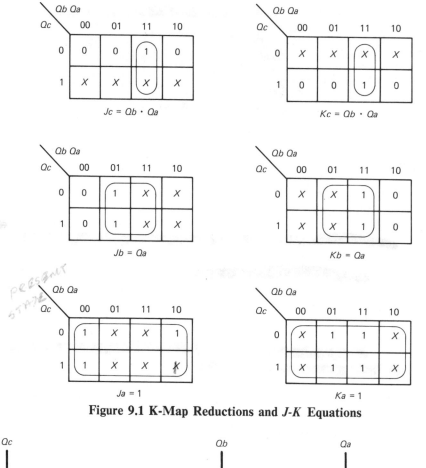

Figure 9.1 K-Map Reductions and *J-K* Equations

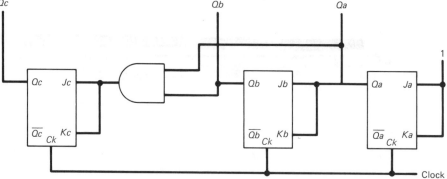

Figure 9.2 3-Bit Synchronous Counter

loading of counters and registers is covered in Chapter 10. Some counters have control inputs that allow selection of either an ascending or descending count sequence.

The 74161 and 74163 are 4-bit synchronous counters. Unlike the 7493 4-bit ripple counter, these counters change state on the rising edge of the input clock rather than the falling edge. The 74161 has an asynchronous clear input that can be used to modify the count length in the same manner illustrated in Chapter 8 for ripple counters. Unfortunately this technique causes glitches with synchronous counters also. To get around this problem, the 74163 provides a synchronous clear input. A synchronous clear does not clear the counter until the first clock edge following the activation of the signal. For example, assume that logic has been added so that the clear signal is activated when state 4 is reached. Though the clear signal becomes active just an instant after switching to state 4, the clear operation will not take place until the next rising clock edge.

To illustrate the use of the synchronous clear, assume we need a divide-by-5 counter. This counter has five states, 0 thru 4. As covered in Chapter 8, the count sequence must be modified so that the counter switches from state 4 to state 0 instead of state 4 to state 5. The method covered in Chapter 8 used the asynchronous clear input to reset the counter the instant it switched to state 5. The brief existence of state 5 is, of course, a glitch. With a synchronous clear input, the clear signal is activated when state 4 is reached. But since it is a synchronous clear, the clear operation does not take place until the next rising clock edge. Thus on the next clock edge, the counter is cleared, instead of switching to state 5.

9.3 State Sequencers

As mentioned in Section 9.1, a synchronous counter is simply a state sequencer with a numerically ascending state sequence. Hence the design of a general state sequencer is similar to the design of a synchronous counter, as covered in Section 9.2.

Assume the sequence shown in the state diagram of Figure 9.3 must be generated. Three flip-flops are required to represent the 3 bits of the highest possible state—state 7 in this example. As with the design of the synchronous counter, we begin with the transition table for this sequencer, which is shown in Table 9.3.

As before, the J-K values shown are those required to make each flip-flop perform the proper state transition. Again, since the J-K inputs are a function of each present state and thus of Qc, Qb, and Qa, we can derive expressions for each J and K input as shown in Figure 9.4. The logic diagram to implement this counter is shown in Figure 9.5.

9.3.1 Unused States

For the sequencer defined by the state diagram of Figure 9.3, states 3, 5, and 6 are not shown. Since these states are not part of the sequence, the expressions for J and K can be made simpler by omitting them from the transition table (their presence

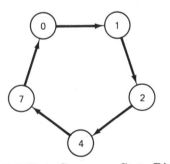

Figure 9.3 State Sequencer State Diagram

Table 9.3 Transition Table for Figure 9.3

Present State			Next State			Flip-Flop Inputs to Perform Change					
Qc	Qb	Qa	Qc	Qb	Qa	Jc	Kc	Jb	Kb	Ja	Ka
0	0	0	0	0	1	0	x	0	x	1	x
0	0	1	0	1	0	0	x	1	x	x	1
0	1	0	1	0	0	1	x	x	1	0	x
1	0	0	1	1	1	x	0	1	x	1	x
1	1	1	0	0	0	x	1	x	1	x	1

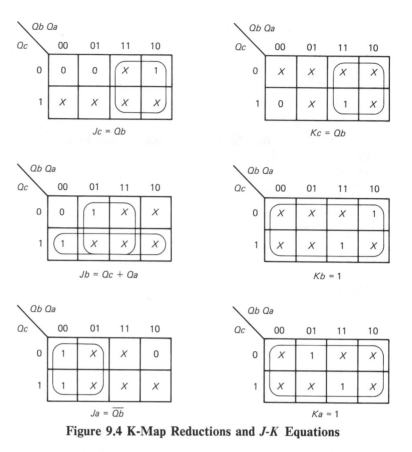

Figure 9.4 K-Map Reductions and *J-K* Equations

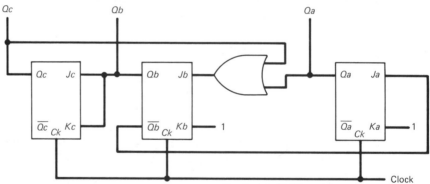

Figure 9.5 Sequencer Logic Diagram for Figure 9.3

results in more don't cares in the K-maps). But what if the sequencer powered up in one of the unused states—what would happen on successive clock cycles? Typically after a design is complete, the sequence from each unused state is determined and appended to the transition table and the state diagram.

The state sequence from an unused state is determined by assuming the present state of the outputs to be one of the unused states. Then by calculating the *J* and *K* inputs to each flip-flop from the equations, the next state of each flip-flop can be determined. Figures 9.6, 9.7, and 9.8 illustrate this process for states 3, 5, and 6, respectively.

As can be seen, state 3 goes to state 4, which is in the normal sequence. State 6 goes to state 0, which is also in the normal sequence. State 5 goes to state 6, which then goes to state 0. These state transitions should be appended to the original state diagram,

	Present state	J inputs	K inputs	Q	Next state
Q_c	0	$J_c = Q_b = 1$	$K_c = Q_b = 1$	Q_c toggles	1
Q_b	1	$J_b = Q_a + Q_c = 1$	$K_b = 1$	Q_b toggles	0
Q_a	1	$J_a = \overline{Q_b} = 0$	$K_a = 1$	Q_a reset	0

Figure 9.6 Next-State Determination from State 3 (011)

	Present state	J inputs	K inputs	Q	Next state
Q_c	1	$J_c = Q_b = 0$	$K_c = Q_b = 0$	Q_c N/C	1
Q_b	0	$J_b = Q_a + Q_c = 1$	$K_b = 1$	Q_b toggles	1
Q_a	1	$J_a = \overline{Q_b} = 1$	$K_a = 1$	Q_a toggles	0

Figure 9.7 Next-State Determination from State 5 (101)

	Present state	J inputs	K inputs	Q	Next state
Q_c	1	$J_c = Q_b = 1$	$K_c = Q_b = 1$	Q_c toggles	0
Q_b	1	$J_b = Q_a + Q_c = 1$	$K_b = 1$	Q_b toggles	0
Q_a	0	$J_a = \overline{Q_b} = 0$	$K_a = 1$	Q_a toggles	0

Figure 9.8 Next-State Determination from State 6 (110)

as shown in Figure 9.9. If an unused state never enters the normal cycle, or if unused states must enter a *specific* state in the cycle, then those state transitions must be included in the transition table of the original design and in the expressions for the *J-K* inputs derived using this extended transition table.

9.3.2 Adding External Control

Often the state sequence generated must be altered by external inputs. In these applications, the next state depends not only on the current state but also on the external inputs. Thus the equations for the *J-K* inputs must be a function not only of the present state but also of the external inputs.

Consider a system with two cycles. One cycle sequences from 0 through 2, and the second cycle sequences from 3 through 5. An external input *Y* is used to determine which cycle to perform. If *Y* is 0, then the first cycle is performed. If *Y* is 1, the second cycle is performed. The first cycle enters and exits only through state 0; the second cycle enters and exits only through state 3. This behavior is shown in the state diagram in Figure 9.10. Shown by each transition vector is the input *Y* determining the transition. Shown in Table 9.4 is the transition table for this system.

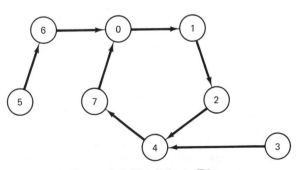

Figure 9.9 Final-State Diagram

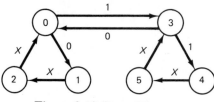

Figure 9.10 State Diagram

Table 9.4 Transition Table for Figure 9.10

Input	Present State			Next State			Flip-Flop Inputs to Perform Change					
Y	Qc	Qb	Qa	Qc	Qb	Qa	Jc	Kc	Jb	Kb	Ja	Ka
0	0	0	0	0	0	1	0	x	0	x	1	x
x	0	0	1	0	1	0	0	x	1	x	x	1
x	0	1	0	0	0	0	0	x	x	1	0	x
1	0	0	0	0	1	1	0	x	1	x	1	x
1	0	1	1	1	0	0	1	x	1	x	1	x
x	1	0	0	1	0	1	x	0	0	x	1	x
x	1	0	1	0	1	1	x	1	1	x	x	0
0	0	1	1	0	0	0	0	x	x	1	x	1

$Jc = Y \cdot Qb \cdot Qa$

$Kc = Qa$

$Jb = Y \cdot \overline{Qc} + Qa$

$Kb = 1$

$Ja = \overline{Qb} + Qa$

$Kc = \overline{Qc}$

Figure 9.11 K-Map Reductions for *J-K* Equations

As mentioned previously, the *J-K* inputs are a function of both the present state (Qc, Qb, Qa), and the input *Y*. We therefore use four input K-maps to determine the reduced expressions for the *J-K* inputs, as shown in Figure 9.11.

The design of the logic diagram and the determination of unused state transitions are left to the reader.

Lab Exercise

Objective

This lab exposes students to a commercially available synchronous counter and lets them determine how to modify its count sequence. The latter part of the lab requires the design of a two-cycle state sequencer with *J-K* flip-flops.

Procedure

1. Using a 74163 synchronous counter, modify the count sequence to run from state 0 through state 6, and then repeat. Produce a logic diagram for your circuit and verify its operation on the logic designer by monitoring the outputs on the LEDs.
2. Design a sequencer using *J-K* flip-flops that has two cycles. The even cycle sequences 0, 2, 4, 6, and repeats. The odd cycle sequences 1, 3, 5, 7, and repeats. An external input selects which cycle to perform. An input of 0 selects the even cycle; an input of 1 selects the odd cycle. The cycles can be entered and exited only through states 0 and 1. Note the similarity to the example in Section 9.3.2. For this sequencer, produce the following:
   (a) A state diagram
   (b) A transition table
   (c) K-maps and reduced expressions for all *J-K* inputs
   (d) A logic diagram for the sequencer
   Wire and test your design.

Have the lab instructor check your work after each circuit is working.

Review Questions

9.1 Produce a transition table for a *D* flip-flop.

9.2 Design a sequencer using *J-K* flip-flops that sequences from 5 down to 0 and then repeats. Produce the following:
   (a) A state diagram
   (b) A transition table
   (c) K-maps and reduced expressions for all *J-K* inputs
   (d) A logic diagram for the sequencer
   (e) An unused-state transition sequence
   (f) A revised-state diagram showing unused states

9.3 Design the sequencer of Question 9.2 using *D* flip-flops.

9.4 Design a sequencer using *J-K* flip-flops that will sequence through the following states and then repeat in response to successive clock transitions:

0, 1, 2, 5, 6, 3, 0 . . .

9.5 Design the sequencer of Question 9.4 using *D* flip-flops.

9.6 If you had to implement a counter with the sequence:

0, 1, 0, 2, 0 . . .

what problems do you foresee that could not be dealt with using a straightforward sequencer design procedure? Suggest how these problems might be resolved. Implement a design using *D* flip-flops.

9.7 A circuit is needed to control an array of eight LEDs, as shown in the accompanying figure. A special lighting sequence is desired such that on successive clock cycles the following LEDs are lit one at a time:

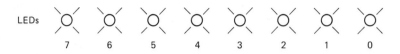

LEDs   7   6   5   4   3   2   1   0

Using a sequencer made from *J-K* flip-flops as the heart of this circuit, complete a design showing the transition table, K-maps, *J-K* equations, and a final logic diagram showing all additional logic required.

9.8 Describe some advantages of a synchronous counter over an asynchronous counter. What are some advantages of asynchronous over synchronous counters?

9.9 Design an up-down 2-bit synchronous counter. The counter should have a mode-control input M. When M is 1, the counter should count up: 0, 1, 2, 3, 0 . . . When the M is 0, the counter should count down: 3, 2, 1, 0, 3 . . . Use *J-K* flip-flops in your design showing transition table, K-maps, final *J-K* equations, and a logic diagram.

9.10 Using *J-K* flip-flops and additional gating as necessary, design a sequential circuit with an input X and an output Z such that Z is 1 only if X has been 1 for three or more consecutive clock cycles. Show the transition table used, K-maps, final *J-K* equations, and a logic diagram. (*Hint:* You must internally count three clock pulses if X is 1. If X is 0, it should not count.)

9.11 Using *J-K* flip-flops and additional gating as necessary, design a sequential circuit with two inputs, X and Y, and an output Z such that Z is 1 only if X and Y have been equal for three or more consecutive clock cycles. Show how this circuit can be made by the addition of a few gates to the circuit of Question 9.10.

Registers

Objectives

1. An in-depth analysis of register design and function.
2. An introduction to the application and interfacing of registers.

10.1 Registers Defined

In general, a *register* is a sequential circuit that can be set to a specific state and retain that state until externally changed. This state might be a multibit binary value in a computer, an address on an address bus, or even a single bit representing an LED that should be on or off. There are many variations and applications of registers, several of which are covered in this chapter.

10.2 Register Design

The *S-R* latch discussed in Chapter 7 is actually a simple 1-bit register. With the (*S*)et input, the latch output can be set to 1. With the (*R*)eset input, the latch output can be set to 0. In both cases, the set or reset output remains until externally changed.

Generally it is desirable to have a single data line rather than separate set and reset lines, as with the *S-R* latch. For example, a *D* flip-flop can be used as a 1-bit register. If the *D* input is 0, the output is set to 0 when the flip-flop is clocked. If the *D* input is 1, the output is set to 1 when the flip-flop is clocked. As with the *S-R* latch, the output remains at 1 or 0 until a new value is loaded into the flip-flop.

With the 7474 Quad *D* flip-flop used in the lab, the output changes state on the rising edge of the clock input. Since the clock input determines when the flip-flop changes state, it can be considered a *load* signal. On the rising edge of this *load* signal, the data on the *D* input is stored in the flip-flop.

10.2.1 Multibit Registers

One-bit registers obviously have limited application. How is a multibit register constructed? Shown in Figure 10.1 is a 4-bit register made from four *D* flip-flops. Each flip-flop stores a single bit. The clock input to each flip-flop comes from a common load signal. On the rising edge of the load signal, each flip-flop stores the value on its *D* input. Registers with more bits are easily constructed by adding an additional flip-flop for each additional bit.

10.2.2 Shift Registers

Data is loaded in parallel into the multibit registers discussed previously, as can be seen in Figure 10.1. But in many applications, the data is not available in a parallel

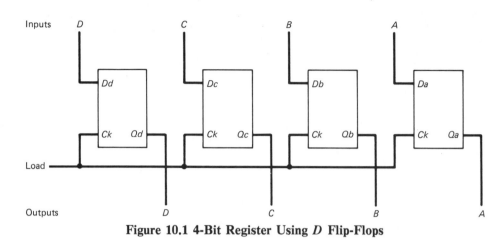

Figure 10.1 4-Bit Register Using *D* Flip-Flops

format. For example, consider the problem of receiving a 4-bit value serially. In this application, all data comes in on a single line, 1 bit at a time. Assume this data is sent the least significant bit first and that a clock signal goes high when each bit is available. The receiving circuit needs to perform a serial-to-parallel conversion, storing each bit as it is received and then somehow presenting the resulting 4-bit value in parallel.

Shown in Figure 10.2 is a circuit that will perform this serial-to-parallel conversion. As can be seen, the *D* input to the first flip-flop is the serial in-data line. The data input to each additional flip-flop is the output of the previous flip-flop. The *bit-available* signal runs to the clock input on each flip-flop. When this signal goes high, each flip-flop will load the data on its *D* input. Because of the connections, the first flip-flop will store the value on the serial input and each additional flip-flop will store the output of the preceding flip-flop. Thus the 4-bit value represented by the four flip-flops is shifted one position to the right, and the leftmost position is filled with the new serial in-data bit. If this operation is repeated four times, the first bit received will be available on *Qa*, the second bit on *Qb*, and so on.

The circuit shown in Figure 10.2 is actually just a 4-bit shift register. On a standard shift register, the *bit available* input is typically referred to as the *clock* input. The shift register shown in Figure 10.2 is a serial-in parallel-out shift register, since data is loaded serially and is available in parallel on the outputs.

A serial-in parallel-out shift register can perform the serial-to-parallel conversion necessary for receiving serial information; but how is the parallel-to-serial conversion necessary for sending serial data performed? In this application, the data is stored in the register in parallel and then shifted out 1 bit at a time. This is accomplished with a parallel-in serial-out shift register.

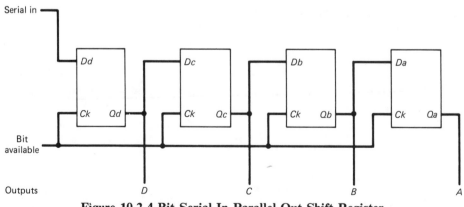

Figure 10.2 4-Bit Serial-In Parallel-Out Shift Register

In general, a shift operation is performed by connecting the input of each flip-flop to the output of the previous flip-flop, as shown in Figure 10.2. But with a parallel-in serial-out shift register, we must also be able to place an external value on the D input to each flip-flop to parallel load the register. Thus each D input can come from one of two sources: the output of the previous flip-flop to perform a shift, or an external data input to parallel load the register. Selecting one of the two sources is easily accomplished using a 2-to-1 mux, as covered in Chapter 5. Because of this selection requirement, a second control input in addition to the clock is required to determine whether to shift or parallel load. A parallel-in serial-out shift register is shown in Figure 10.3. The selection between shift and parallel load is accomplished with the $\overline{\text{SHIFT}}$/LOAD line. This terminology indicates that a shift is performed if the line is low and a parallel load performed if the line is high. Assuming 7474 Quad D flip-flops as used in the lab, the actual shift or load operation takes place on the rising edge of the clock signal. The serial output of the shift register is on Qa. Since something must be shifted into the first flip-flop on a shift operation, a parallel-in shift register will generally also have a serial input, as shown in Figure 10.3.

To use this shift register to serially transmit a 4-bit word, the $\overline{\text{SHIFT}}$/LOAD line is initially set high and the 4-bit value loaded into the register on the first clock cycle. At this time, the least significant bit is available at the output. Before the next clock cycle, the $\overline{\text{SHIFT}}$/LOAD line is returned low so that on successive clock cycles, the 4-bit value is shifted right and the next bit is available at the output.

10.2.3 Transparent Latches

With the registers discussed so far, values are stored based on a clock or load signal. The outputs change only when a new value is loaded on the clock edge. Another type of register, typically referred to as a *transparent latch,* operates in a slightly different manner. Instead of a clock or load signal, the transparent latch has a gate or enable input. When enabled, the outputs of the latch directly follow the inputs of the latch. Except for the propagation delay of the latch, the latch is transparent. When not enabled, the outputs remain in the same state they were in the instant the enable signal went off. For example, assume a transparent latch with an active-high enable. While this enable is high, the outputs follow the inputs. When the enable line returns low, the outputs remain in the same state they were in the instant before the enable went low. Note the similarity to the *S-R* gated latch discussed in Section 7.3.1. A multibit transparent latch is easily made from multiple-gated *S-R* latches (see Review Question 10.3).

Transparent latches have several applications. For example, to reduce the number

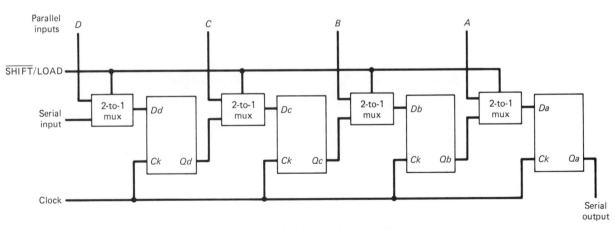

Figure 10.3 4-Bit Parallel-In Serial-Out Shift Register

of pins on the chip, some microprocessors multiplex the data lines with the address lines —the same pins are used for both the data bus and the address bus. During one-half of a master clock cycle, the pins represent the address being accessed. During the second half of the master clock cycle, those same pins represent the data lines. With this type of processor, a transparent latch can be used to maintain the original value of the address lines during the second half of the master clock cycle when those lines turn into data lines. Specifically, during the first half of the master clock cycle, the latch is enabled and passes the address lines through to its outputs. During the second half of the clock cycle, the latch is disabled and maintains the original address-line values on its outputs.

10.3 Register Interfacing

We have considered the design and operation of registers, but how are these registers incorporated into a digital circuit? To illustrate applications and interfacing techniques, we will interface a register with a hypothetical microprocessor system. Details of microprocessor address decoding and timing are beyond the scope of this text; thus these will be treated in a simplified manner.

Assume this microprocessor has a 16-bit address bus for addressing memory and peripherals. In Chapter 5, a simple address-decoding scheme was illustrated using a decoder to enable one of four RAM chips. Based on the address, the decoder set the enable line for the proper RAM chip low. We will assume a similar address-decoding scheme to interface a register to the microprocessor. When the microprocessor addresses the register, an enable line for the register will go low. We will refer to this line as $\overline{\text{ENABLE}}$. An 8-bit data bus is used to transfer data to and from the microprocessor. A $\overline{\text{READ}}$ line goes low when the microprocessor reads from the addressed device, and a $\overline{\text{WRITE}}$ line goes low when the microprocessor writes to the addressed device. To synchronize events, there is a master clock signal, MCLK. Addresses (and hence $\overline{\text{ENABLE}}$) and the $\overline{\text{READ}}$ and $\overline{\text{WRITE}}$ signals become valid during the low half of the MCLK cycle. The microprocessor reads data from the data bus on the falling edge of MCLK. Data written from the microprocessor is valid on the falling edge of MCLK. Therefore, peripherals should store data from the data bus on the falling edge of MCLK. This timing information is shown in Figure 10.4.

10.3.1 Interfacing a Write-Only Register

An external register has many uses in a microprocessor system. For example, the outputs of an 8-bit register might be used to run eight indicator LEDs. The microprocessor can light any combination of LEDs by storing different values in the register. Here the microprocessor only writes to the register. How is the register interfaced with the microprocessor in this type of write-only application?

The primary concern is when to load the register with the data on the data bus. Obviously this should be done only when the microprocessor is writing to the register. When this occurs, the $\overline{\text{WRITE}}$ line goes low, and the $\overline{\text{ENABLE}}$ line for the register goes low since the address on the address bus is that of the register. If both the $\overline{\text{WRITE}}$ and $\overline{\text{ENABLE}}$ lines are low, then on the falling edge of MCLK, the register should be loaded with the data on the data bus. The registers designed in Section 10.2.1 loaded

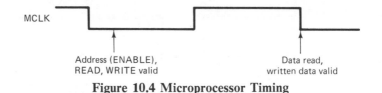

MCLK

Address (ENABLE),
READ, WRITE valid

Data read,
written data valid

Figure 10.4 Microprocessor Timing

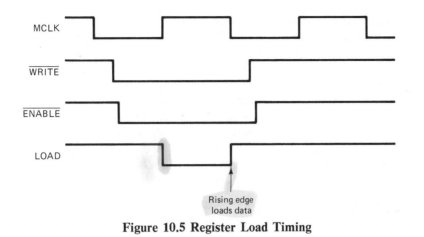

Figure 10.5 Register Load Timing

data on the rising edge of their clock or load input. We will refer to this signal as LOAD. The interface requirements can then be summarized as follows: If $\overline{\text{WRITE}}$ is low and $\overline{\text{ENABLE}}$ is low, then on the falling edge of MCLK, LOAD should go high to load the register from the data bus. This sequence could occur as shown in the timing diagram in Figure 10.5.

Now that the interface timing requirements are defined, how can the LOAD signal shown be generated? From Figure 10.5 we see that if $\overline{\text{WRITE}}$ is low and $\overline{\text{ENABLE}}$ is low and MCLK is high, then the LOAD line should go low. This is simply the NAND function, as illustrated in Figure 10.6. Inspecting this circuit, we see the LOAD line goes low only when $\overline{\text{WRITE}}$ and $\overline{\text{ENABLE}}$ are both low and MCLK is high. As soon as MCLK returns low, LOAD goes high to load the register as desired.

10.3.2 Interfacing a Read-Only Register

Other uses of registers include applications where an external event stores a value in the register to be read later by the microprocessor. For example, consider a keyboard-encoding circuit for a simple numeric keypad. This circuit converts a key closure into a binary value representing the key pressed, stores this value in a register, and then signals the microprocessor that a new key value is available. The microprocessor then reads the register to get the value of the key pressed.

Interface requirements for reading a register are quite different than those for writing a register. In Section 10.3.1 we illustrated how to generate a LOAD signal to load the register at the proper time. How can we generate a signal to read the register at the proper time? Recall that the outputs of the register are simply the outputs of flip-flops. These outputs are always on. If a flip-flop is storing a 1, the output will be 1. How can we selectively read this register if the outputs are always on?

To answer this question, we look at the operation of the data bus. When the microprocessor is writing, its outputs are driving the data bus—no other device is attempting to put a value on the data bus. Similarly, when the microprocessor is reading, only the outputs of the addressed device are driving the data bus. The only

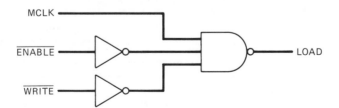

Figure 10.6 Generation of Load Signal

time that the outputs of a device should be *connected* to the data bus is when the microprocessor is reading that device. At all other times, a device's outputs should be effectively disconnected from the data bus. The obvious question is how can the outputs be *disconnected* from the data bus?

Recall from Chapter 1 the existence of gates with three-state outputs. These outputs can assume the normal 1 or 0 states, but in addition they can assume a third state that appears to be an open circuit—as if nothing is there. When the output is in this third state, it is effectively disconnected from the circuit. To control this third state, an enable line is used to enable or disable the output. When enabled, the output assumes its normal 1 or 0 state. When not enabled, the output is in the third state—disconnected. Both active-low and active-high enables are available. With the active-low enable, the output is enabled when the enable line is low. With the active-high enable, the output is enabled when the enable line is high.

Shown in Figure 10.7 is a 4-bit register with four tristate gates at the outputs. These tristate gates simply pass the input value to the output with the capability of enabling or disabling the output with an active-low enable, as previously mentioned. These gates are often referred to as tristate buffers. With the enable line low, the outputs of the register are passed through the tristate buffers to the final output. When the enable line is high, there appears to be nothing connected to the final outputs.

These buffers are available on TTL chips. For example, the 74367 has six of these buffers on a single chip. The 74368 contains six tristate inverters. The 74LS244 has eight of these buffers on a single chip and is especially useful in typical 8-bit microprocessor applications.

Using tristate buffers we can interface the output of a register to a data bus. The outputs should be connected to the data bus only when the microprocessor is reading from the register. When this occurs, the $\overline{\text{READ}}$ line goes low and the $\overline{\text{ENABLE}}$ line for the register goes low since the address on the address bus is that of the register. The outputs of the register must be connected to the data bus before the falling edge of MCLK when the processor reads the value on the bus. To connect the outputs of the register to the data bus, we simply enable the outputs of the tristate buffers. We will assume an active-low output-enable line as used in Figure 10.7. The proper sequence of events could occur as shown in the timing diagram in Figure 10.8. As can be seen, the output enable goes low only when $\overline{\text{READ}}$ and $\overline{\text{ENABLE}}$ are both low and MCLK is high. During this time, the outputs of the register are connected to the data bus and will therefore be valid by the falling edge of MCLK. Generation of the output-enable

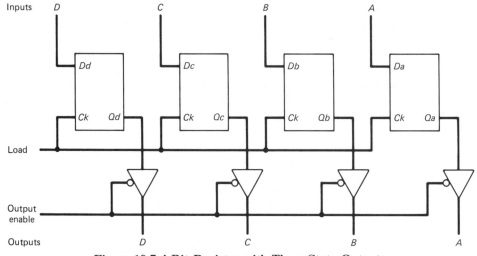

Figure 10.7 4-Bit Register with Three-State Outputs

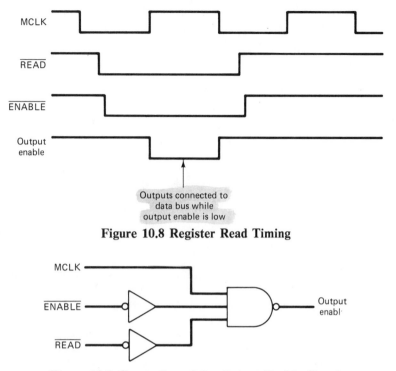

Figure 10.8 Register Read Timing

Figure 10.9 Generation of the Output-Enable Signal

signal is just a NAND operation similar to the generation of the LOAD signal. A circuit to generate the output enable is shown in Figure 10.9.

10.3.3 Interfacing a Read/Write Register

Some applications use a register as a fast temporary memory cell; the general-purpose register in a computer is one example. In these applications the register is both read and written. The same address is used both to read and write the register, the READ and WRITE lines determining which operation is to take place. Interface requirements for a write operation and for a read operation have already been defined in Sections 10.3.1 and 10.3.2. By combining the two circuits shown in Figures 10.6 and 10.9, we can generate the proper LOAD- and output-enable signals.

10.4 Commercially Available Registers

Many variations of registers are available on TTL chips. The 74175 is a 4-bit register using four D flip-flops like the circuit shown in Figure 10.1. The 74174 is similar to the 74175; but it has 6 bits, while the 74273 has 8 bits. In addition to the data lines and a clock input, each of these registers has a clear input to initialize the register to all 0s.

To reduce interface requirements, the 74LS377 8-bit register has both an enable and clock input for storing data. Unless the enable line is low, the clock input is ignored. The 74LS364 8-bit register has built in tristate outputs, eliminating the need for a separate tristate buffer chip. The 74LS363 is similar to the 74LS364, but the '363 is a transparent latch. Instead of a clock input, it has an enable input, as described in Section 10.2.3.

Shift registers are available in virtually countless variety. The 7495 is a 4-bit parallel-in parallel-out shift register. This register can be parallel loaded or serially shifted in the same way as the shift register shown in Figure 10.3. In addition, the output of each flip-flop is brought to the outside—thus providing the parallel-out capability. The 74164 is an 8-bit serial-in parallel-out shift register similar to the 4-bit version shown in Figure 10.2. The parallel-in serial-out counterpart of the 74164 is the 74166, which is similar to the 4-bit register shown in Figure 10.3.

Lab Exercise

Objective

This lab exposes the student to the use of various register configurations. First, a register and shift register are built from individual *D* flip-flops, and then a 7495 shift register is used to perform the same function. Finally, 4-bit serial communication is implemented using two 7495 shift registers. This problem leaves many design and operation details to the student, requiring in-depth study.

Procedure

1. Using 7474 *D* flip-flops, wire a 4-bit register similar to the one in Figure 10.1. Connect the inputs to switches on the logic designer and the outputs to the LEDs. Using a pulser switch as the clock input to load data, store several values in the register. Does the output change when the inputs change or on the clock edge?
2. Using 7474 *D* flip-flops, wire a 4-bit serial-in parallel-out shift register similar to Figure 10.2. Connect the input to a switch on the logic designer and the outputs to the LEDs. Connect the clock input to a pulser switch. Data is loaded 1 bit at a time by setting the switch to the desired value and pressing the pulser to load the value into the most significant bit of the register. Repeating this four times loads a 4-bit value into the register.
3. The 7495 shift register can perform both serial-in parallel-out and parallel-in serial-out operations. Noting the use of the mode control and clock inputs from the data sheet for the 7495, use the 7495 to perform the same function as in (2).
4. Use two 7495s to perform serial communication. One is used to parallel-load data from the switches and send the value serially, while the second is used to serially receive the value and display it in parallel on the LEDs. Note carefully the transmit-receive sequence as outlined in Section 10.2.2. Produce a logic diagram and wire and test your circuit. (*Hint:* To clock the two registers, connect the output of a pulser to the clock(s) of the transmit register and then invert this signal for the clock of the receive register. This way, the transmit register is loaded or shifted when the pulser is pushed, and the receiving register loads the data when the pulser is released. Note carefully on which clock edge the 7495 triggers, and from this insure that the transmit register gets the proper clock edge first.)

Have the lab instructor check your work after each circuit is working.

Review Questions

10.1 Design a 4-bit register using *J-K* flip-flops.
10.2 In Figure 10.6, 2-to-1 muxes were used to select either a shift- or parallel-load operation. Show the design at the gate level of one of these muxes, labeling the serial input and the parallel input such that the control line provides the $\overline{\text{SHIFT}}$/LOAD operation discussed in Section 10.2.2.
10.3 Design a 2-bit transparent latch at the gate level. This latch should have a single data-input line for each bit and an active-high enable.
10.4 The 7476 *J-K* flip-flop used in the lab changes state on the falling edge of the clock. Show the generation of the load signal as discussed in Section 10.3.1 for a register made from 7476 *J-K* flip-flops.
10.5 Some microprocessors have a single R/$\overline{\text{W}}$ line rather than separate $\overline{\text{READ}}$ and $\overline{\text{WRITE}}$ lines. This line is high if a read operation is to take place and low if a write operation is to take place. Show the generation of the load signal for a rising edge-triggered register if an R/$\overline{\text{W}}$ line is used rather than $\overline{\text{READ}}$ and $\overline{\text{WRITE}}$.

10.6 Show the generation of the output enable for an active-low enable tristate buffer in a system using an R/$\overline{\text{W}}$ line, as described in Question 10.3.

10.7 In Chapter 5, muxes were used to select among multiple-input sources. When selecting among multiple-bit words, a mux was used for each bit of the word. Show how a 4-bit word can be selected from one of four sources using tristate buffers and a decoder. This function of this circuit is shown in the accompanying illustration:

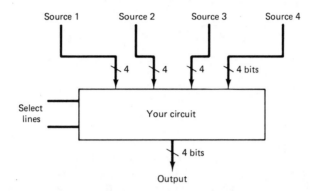

10.8 Using the microprocessor characteristics described in Section 10.3.1, show the interface circuitry for a 74LS377 register as described in Section 10.4 for a write operation. The '377 loads data on the rising edge of its clock when enabled.

Boolean Algebra

In 1854 an English mathematician, George Boole, composed the treatise "An Investigation of the Laws of Thought on Which to Found the Mathematical Theories of Logic and Probabilities," which was to perform a mathematical analysis of logic. As a result of his investigation and the construction of a "logical algebra," subsequent mathematicians and logicians were led into several new fields of mathematics. From the philosophy to the "maturing" mathematics developed an algebra that is now used in the design of digital computer logical circuitry. This algebra was called *boolean algebra*.

In 1938, a research assistant, Claude E. Shannon, in the Department of Electrical Engineering at Massachusetts Institute of Technology, suggested that boolean algebra be used to solve problems in relay-switching circuit design. The basic techniques suggested by Shannon were later used for the design and analysis of switching circuits used in modern high-speed computers. There are many advantages to the use of mathematical techniques—for example, convenience in reducing and simplifying an expression (with the aid of a computer). Furthermore, boolean algebra is an economical and straightforward way of describing the circuitry used in computers. A knowledge of boolean algebra is therefore indispensable in the computer field.

The following rules apply to logical states.

1. Two states exist; any pair of different conditions, usually "true" and "false," can be chosen.
2. Every quantity must exist in one of the two chosen states; no other value is allowed.
3. Every logical quantity is single valued. No quantity may be simultaneously "true" and "false."
4. Any quantity that is "true" equals any other quantity that is "true." Any "false" quantity is equal to any other "false" quantity.
5. Every quantity has an opposite. If the quantity is "true," then the inverse (complement) is "false"; and vice versa.
6. A logical quantity may be either a constant or a variable.
7. Logical quantities may be physically represented in many ways, such as:
   (a) Electrically (with two different voltages)
   (b) Mechanically (by the position of a toggle switch)
   (c) Optically (by the presence or absence of light)

Since boolean expressions contain variables that can have a value of either a 0 or 1, the result of a boolean function, no matter how complex, can only have a value of 0 or 1. For example, the expression:

$$(A \cdot \overline{B} + (C + \overline{D} + E))$$

can only have a value of 0 or 1.

Boolean Laws

The following laws are useful in understanding and simplifying complex boolean functions:

1. Commutative law

$$A + B = B + A$$
$$A \cdot B = B \cdot A$$

2. Associative law

$$A + B + C = A + (B + C) = (A + B) + C$$
$$A \cdot B \cdot C = A \cdot (B \cdot C) = (A \cdot B) \cdot C$$

3. Distributive law

$$A \cdot (B + C) = A \cdot B + A \cdot C$$
$$A + (B \cdot C) = (A + B) \cdot (A + C)$$

Postulates of Boolean Algebra

$$A + \overline{A} = 1$$
$$A \cdot \overline{A} = 0$$
$$A \cdot A = A$$
$$\overline{A} + \overline{A} = \overline{A}$$
$$A + B = B + A$$
$$\overline{(\overline{A})} = A$$
$$A \cdot 0 = 0$$
$$A \cdot 1 = A$$
$$A + 0 = A$$
$$A + 1 = 1$$
$$A + A = A$$
$$A + AB = A$$
$$A + \overline{A}B = A + B$$
$$(A + B) \cdot (A + C) = A + BC$$
$$(A + B)B = AB$$
$$AB + AC = A \cdot (B + C)$$
$$AB = BA$$
$$\overline{(A + B)} = \overline{A} \cdot \overline{B} \ \} \text{ DeMorgan's theorem}$$
$$\overline{(AB)} = \overline{A} + \overline{B} \ \} \text{ DeMorgan's theorem}$$

TTL Parts and Layouts

This appendix is composed of extracts from Texas Instruments' *TTL Data Book for Design Engineers* (2d ed., 1976).

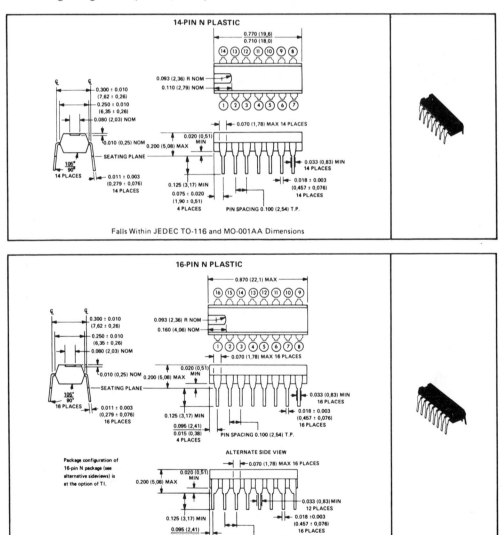

N plastic dual-in-line packages

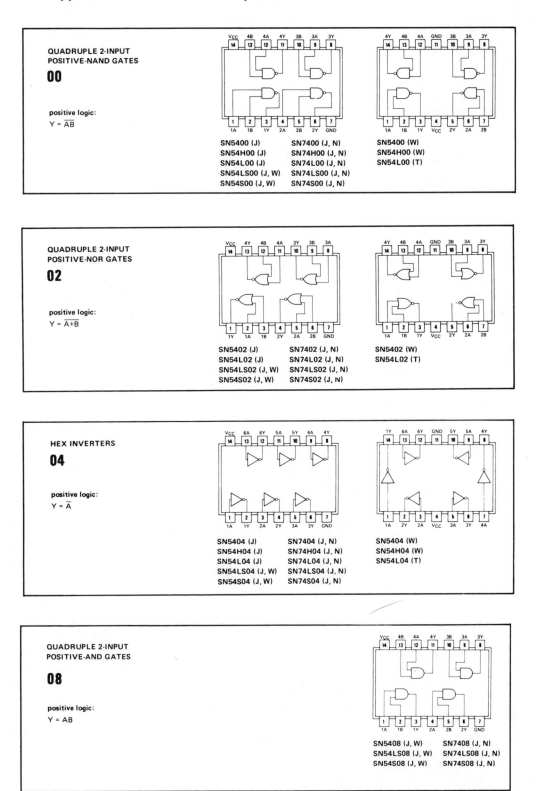

QUADRUPLE 2-INPUT POSITIVE-NAND GATES

00

positive logic:
$Y = \overline{AB}$

SN5400 (J)	SN7400 (J, N)
SN54H00 (J)	SN74H00 (J, N)
SN54L00 (J)	SN74L00 (J, N)
SN54LS00 (J, W)	SN74LS00 (J, N)
SN54S00 (J, W)	SN74S00 (J, N)

SN5400 (W)
SN54H00 (W)
SN54L00 (T)

QUADRUPLE 2-INPUT POSITIVE-NOR GATES

02

positive logic:
$Y = \overline{A+B}$

SN5402 (J)	SN7402 (J, N)
SN54L02 (J)	SN74L02 (J, N)
SN54LS02 (J, W)	SN74LS02 (J, N)
SN54S02 (J, W)	SN74S02 (J, N)

SN5402 (W)
SN54L02 (T)

HEX INVERTERS

04

positive logic:
$Y = \overline{A}$

SN5404 (J)	SN7404 (J, N)
SN54H04 (J)	SN74H04 (J, N)
SN54L04 (J)	SN74L04 (J, N)
SN54LS04 (J, W)	SN74LS04 (J, N)
SN54S04 (J, W)	SN74S04 (J, N)

SN5404 (W)
SN54H04 (W)
SN54L04 (T)

QUADRUPLE 2-INPUT POSITIVE-AND GATES

08

positive logic:
$Y = AB$

SN5408 (J, W)	SN7408 (J, N)
SN54LS08 (J, W)	SN74LS08 (J, N)
SN54S08 (J, W)	SN74S08 (J, N)

**TRIPLE 3-INPUT
POSITIVE-NAND GATES**

10

positive logic:

$Y = \overline{ABC}$

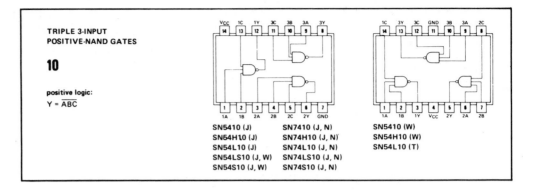

SN5410 (J)	SN7410 (J, N)
SN54H10 (J)	SN74H10 (J, N)
SN54L10 (J)	SN74L10 (J, N)
SN54LS10 (J, W)	SN74LS10 (J, N)
SN54S10 (J, W)	SN74S10 (J, N)

SN5410 (W)
SN54H10 (W)
SN54L10 (T)

**DUAL 4-INPUT
POSITIVE-NAND GATES**

20

positive logic:

$Y = \overline{ABCD}$

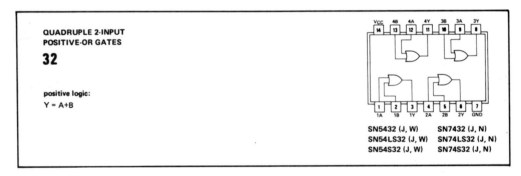

SN5420 (J)	SN7420 (J, N)
SN54H20 (J)	SN74H20 (J, N)
SN54L20 (J)	SN74L20 (J, N)
SN54LS20 (J, W)	SN74LS20 (J, N)
SN54S20 (J, W)	SN74S20 (J, N)

SN5420 (W)
SN54H20 (W)
SN54L20 (T)

NC—No internal connection

**QUADRUPLE 2-INPUT
POSITIVE-OR GATES**

32

positive logic:

$Y = A+B$

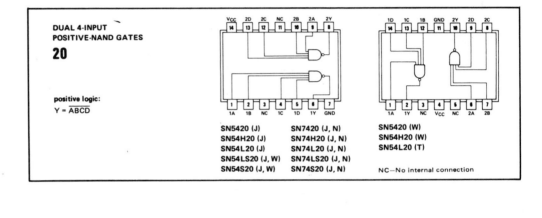

SN5432 (J, W)	SN7432 (J, N)
SN54LS32 (J, W)	SN74LS32 (J, N)
SN54S32 (J, W)	SN74S32 (J, N)

TYPES SN5446A, '47A, '48, '49, SN54L46, 'L47, SN54LS47, 'LS48, 'LS49, SN7446A, '47A, '48, SN74L46, 'L47, SN74LS47, 'LS48, 'LS49
BCD-TO-SEVEN-SEGMENT DECODERS/DRIVERS
BULLETIN NO. DL-S 7611811, MARCH 1974—REVISED OCTOBER 1976

'46A, '47A, 'L46, 'L47, 'LS47 feature	'48, 'LS48 feature	'49, 'LS49 feature
• Open-Collector Outputs Drive Indicators Directly	• Internal Pull-Ups Eliminate Need for External Resistors	• Open-Collector Outputs
• Lamp-Test Provision	• Lamp-Test Provision	• Blanking Input
• Leading/Trailing Zero Suppression	• Leading/Trailing Zero Suppression	

• All Circuit Types Feature Lamp Intensity Modulation Capability

TYPE	DRIVER OUTPUTS				TYPICAL POWER DISSIPATION	PACKAGES
	ACTIVE LEVEL	OUTPUT CONFIGURATION	SINK CURRENT	MAX VOLTAGE		
SN5446A	low	open-collector	40 mA	30 V	320 mW	J, W
SN5447A	low	open-collector	40 mA	15 V	320 mW	J, W
SN5448	high	2-kΩ pull-up	6.4 mA	5.5 V	265 mW	J, W
SN5449	high	open-collector	10 mA	5.5 V	165 mW	W
SN54L46	low	open-collector	20 mA	30 V	160 mW	J
SN54L47	low	open-collector	20 mA	15 V	160 mW	J
SN54LS47	low	open-collector	12 mA	15 V	35 mW	J, W
SN54LS48	high	2-kΩ pull-up	2 mA	5.5 V	125 mW	J, W
SN54LS49	high	open-collector	4 mA	5.5 V	40 mW	J, W
SN7446A	low	open-collector	40 mA	30 V	320 mW	J, N
SN7447A	low	open-collector	40 mA	15 V	320 mW	J, N
SN7448	high	2-kΩ pull-up	6.4 mA	5.5 V	265 mW	J, N
SN74L46	low	open-collector	20 mA	30 V	160 mW	J, N
SN74L47	low	open-collector	20 mA	15 V	160 mW	J, N
SN74LS47	low	open-collector	24 mA	15 V	35 mW	J, N
SN74LS48	high	2-kΩ pull-up	6 mA	5.5 V	125 mW	J, N
SN74LS49	high	open-collector	8 mA	5.5 V	40 mW	J, N

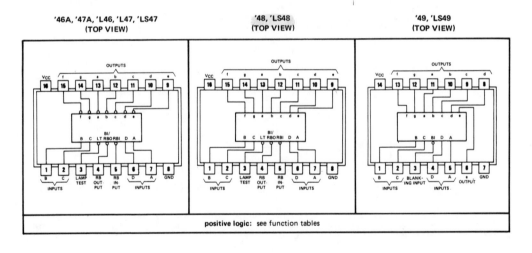

'46A, '47A, 'L46, 'L47, 'LS47 (TOP VIEW)   '48, 'LS48 (TOP VIEW)   '49, 'LS49 (TOP VIEW)

positive logic: see function tables

TYPES SN5446A, '47A, '48, '49, SN54L46, 'L47, SN54LS47, 'LS48, 'LS49, SN7446A, '47A, '48, SN74L46, 'L47, SN74LS47, 'LS48, 'LS49 BCD-TO-SEVEN-SEGMENT DECODERS/DRIVERS

'48, 'LS48 FUNCTION TABLE

DECIMAL OR FUNCTION	INPUTS						BI/RBO†	OUTPUTS							NOTE
	LT	RBI	D	C	B	A		a	b	c	d	e	f	g	
0	H	H	L	L	L	L	H	H	H	H	H	H	H	L	
1	H	X	L	L	L	H	H	L	H	H	L	L	L	L	
2	H	X	L	L	H	L	H	H	H	L	H	H	L	H	
3	H	X	L	L	H	H	H	H	H	H	H	L	L	H	
4	H	X	L	H	L	L	H	L	H	H	L	L	H	H	
5	H	X	L	H	L	H	H	H	L	H	H	L	H	H	
6	H	X	L	H	H	L	H	L	L	H	H	H	H	H	
7	H	X	L	H	H	H	H	H	H	H	L	L	L	L	
8	H	X	H	L	L	L	H	H	H	H	H	H	H	H	1
9	H	X	H	L	L	H	H	H	H	H	L	L	H	H	
10	H	X	H	L	H	L	H	L	L	L	H	H	L	H	
11	H	X	H	L	H	H	H	L	L	H	H	L	L	H	
12	H	X	H	H	L	L	H	L	H	L	L	L	H	H	
13	H	X	H	H	L	H	H	H	L	L	H	L	H	H	
14	H	X	H	H	H	L	H	L	L	L	H	H	H	H	
15	H	X	H	H	H	H	H	L	L	L	L	L	L	L	
BI	X	X	X	X	X	X	L	L	L	L	L	L	L	L	2
RBI	H	L	L	L	L	L	L	L	L	L	L	L	L	L	3
LT	L	X	X	X	X	X	H	H	H	H	H	H	H	H	4

H = high level, L = low level, X = irrelevant

NOTES: 1. The blanking input (BI) must be open or held at a high logic level when output functions 0 through 15 are desired. The ripple-blanking input (RBI) must be open or high, if blanking of a decimal zero is not desired.
2. When a low logic level is applied directly to the blanking input (BI), all segment outputs are low regardless of the level of any other input.
3. When ripple-blanking input (RBI) and inputs A, B, C, and D are at a low level with the lamp-test input high, all segment outputs go low and the ripple-blanking output (RBO) goes to a low level (response condition).
4. When the blanking input/ripple-blanking output (BI/RBO) is open or held high and a low is applied to the lamp-test input, all segment outputs are high.

†BI/RBO is wire-AND logic serving as blanking input (BI) and/or ripple-blanking output (RBO).

'49, 'LS49 FUNCTION TABLE

DECIMAL OR FUNCTION	INPUTS					OUTPUTS							NOTE
	D	C	B	A	BI	a	b	c	d	e	f	g	
0	L	L	L	L	H	H	H	H	H	H	H	L	
1	L	L	L	H	H	L	H	H	L	L	L	L	
2	L	L	H	L	H	H	H	L	H	H	L	H	
3	L	L	H	H	H	H	H	H	H	L	L	H	
4	L	H	L	L	H	L	H	H	L	L	H	H	
5	L	H	L	H	H	H	L	H	H	L	H	H	
6	L	H	H	L	H	L	L	H	H	H	H	H	
7	L	H	H	H	H	H	H	H	L	L	L	L	
8	H	L	L	L	H	H	H	H	H	H	H	H	1
9	H	L	L	H	H	H	H	H	L	L	H	H	
10	H	L	H	L	H	L	L	L	H	H	L	H	
11	H	L	H	H	H	L	L	H	H	L	L	H	
12	H	H	L	L	H	L	H	L	L	L	H	H	
13	H	H	L	H	H	H	L	L	H	L	H	H	
14	H	H	H	L	H	L	L	L	H	H	H	H	
15	H	H	H	H	H	L	L	L	L	L	L	L	
BI	X	X	X	X	L	L	L	L	L	L	L	L	2

H = high level, L = low level, X = irrelevant

NOTES: 1. The blanking input (BI) must be open or held at a high logic level when output functions 0 through 15 are desired.
2. When a low logic level is applied directly to the blanking input (BI), all segment outputs are low regardless of the level of any other input.

DUAL D-TYPE POSITIVE-EDGE-TRIGGERED FLIP-FLOPS WITH PRESET AND CLEAR

74

FUNCTION TABLE

INPUTS				OUTPUTS	
PRESET	CLEAR	CLOCK	D	Q	Q̄
L	H	X	X	H	L
H	L	X	X	L	H
L	L	X	X	H*	H*
H	H	↑	H	H	L
H	H	↑	L	L	H
H	H	L	X	Q_0	$\bar{Q}_0$

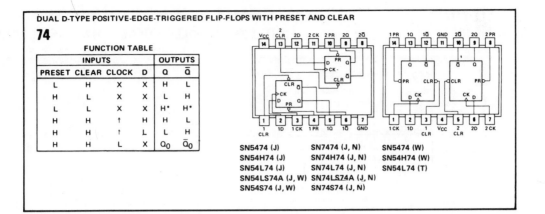

SN5474 (J)   SN7474 (J, N)   SN5474 (W)
SN54H74 (J)   SN74H74 (J, N)   SN54H74 (W)
SN54L74 (J)   SN74L74 (J, N)   SN54L74 (T)
SN54LS74A (J, W) SN74LS74A (J, N)
SN54S74 (J, W) SN74S74 (J, N)

DUAL J-K FLIP-FLOPS WITH PRESET AND CLEAR

76

'76, 'H76 FUNCTION TABLE

INPUTS					OUTPUTS	
PRESET	CLEAR	CLOCK	J	K	Q	Q̄
L	H	X	X	X	H	L
H	L	X	X	X	L	H
L	L	X	X	X	H*	H*
H	H	⊓	L	L	Q_0	$\bar{Q}_0$
H	H	⊓	H	L	H	L
H	H	⊓	L	H	L	H
H	H	⊓	H	H	TOGGLE	

'LS76A FUNCTION TABLE

INPUTS					OUTPUTS	
PRESET	CLEAR	CLOCK	J	K	Q	Q̄
L	H	X	X	X	H	L
H	L	X	X	X	L	H
L	L	X	X	X	H*	H*
H	H	↓	L	L	Q_0	$\bar{Q}_0$
H	H	↓	H	L	H	L
H	H	↓	L	H	L	H
H	H	↓	H	H	TOGGLE	
H	H	H	X	X	Q_0	$\bar{Q}_0$

SN5476 (J, W)   SN7476 (J, N)
SN54H76 (J, W)   SN74H76 (J, N)
SN54LS76A (J, W)   SN74LS76A (J, N)

**TTL
MSI**

TYPES SN5483A, SN54LS83A, SN7483A, SN74LS83A
4-BIT BINARY FULL ADDERS WITH FAST CARRY

BULLETIN NO. DL-S 7611853, MARCH 1974—REVISED OCTOBER 1976

- Full-Carry Look-Ahead across the Four Bits
- Systems Achieve Partial Look-Ahead Performance with the Economy of Ripple Carry
- SN54283/SN74283 and SN54LS283/SN74LS283 Are Recommended For New Designs as They Feature Supply Voltage and Ground on Corner Pins to Simplify Board Layout

TYPE	TYPICAL ADD TIMES		TYPICAL POWER DISSIPATION PER 4-BIT ADDER
	TWO 8-BIT WORDS	TWO 16-BIT WORDS	
'83A	23 ns	43 ns	310 mW
'LS83A	25 ns	45 ns	95 mW

description

These improved full adders perform the addition of two 4-bit binary numbers. The sum (Σ) outputs are provided for each bit and the resultant carry (C4) is obtained from the fourth bit. These adders feature full internal look ahead across all four bits generating the carry term in ten nanoseconds typically. This provides the system designer with partial look-ahead performance at the economy and reduced package count of a ripple-carry implementation.

The adder logic, including the carry, is implemented in its true form meaning that the end-around carry can be accomplished without the need for logic or level inversion.

Designed for medium-speed applications, the circuits utilize transistor-transistor logic that is compatible with most other TTL families and other saturated low-level logic families.

Series 54 and 54LS circuits are characterized for operation over the full military temperature range of $-55°C$ to $125°C$, and Series 74 and 74LS circuits are characterized for operation from $0°C$ to $70°C$.

SN5483A, SN54LS83A . . . J OR W PACKAGE
SN7483A, SN74LS83A . . . J OR N PACKAGE
(TOP VIEW)

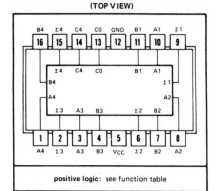

positive logic: see function table

FUNCTION TABLE

INPUT				OUTPUT					
				WHEN C0 = L			WHEN C0 = H		
						WHEN C2 = L			WHEN C2 = H
A1 / A3	B1 / B3	A2 / A4	B2 / B4	Σ1 / Σ3	Σ2 / Σ4	C2 / C4	Σ1 / Σ3	Σ2 / Σ4	C2 / C4
L	L	L	L	L	L	L	H	L	L
H	L	L	L	H	L	L	L	H	L
L	H	L	L	H	L	L	L	H	L
H	H	L	L	L	H	L	H	H	L
L	L	H	L	L	H	L	H	H	L
H	L	H	L	H	H	L	L	L	H
L	H	H	L	H	H	L	L	L	H
H	H	H	L	L	L	H	H	L	H
L	L	L	H	L	H	L	H	H	L
H	L	L	H	H	H	L	L	L	H
L	H	L	H	H	H	L	L	L	H
H	H	L	H	L	L	H	H	L	H
L	L	H	H	L	L	H	H	L	H
H	L	H	H	H	L	H	L	H	H
L	H	H	H	H	L	H	L	H	H
H	H	H	H	L	H	H	H	H	H

H = high level, L = low level

NOTE: Input conditions at A1, B1, A2, B2, and C0 are used to determine outputs $\Sigma1$ and $\Sigma2$ and the value of the internal carry C2. The values at C2, A3, B3, A4, and B4 are then used to determine outputs $\Sigma3$, $\Sigma4$, and C4.

absolute maximum ratings over operating free-air temperature range (unless otherwise noted)

Supply voltage, V_{CC} (see Note 1)	. . . .	7 V
Input voltage: '83A	. . . .	5.5 V
'LS83A	. . . .	7 V
Interemitter voltage (see Note 2)	. . . .	5.5 V
Operating free-air temperature range: SN5483A, SN54LS83A	. . . .	$-55°C$ to $125°C$
SN7483A, SN74LS83A	. . . .	$0°C$ to $70°C$
Storage temperature range	. . . .	$-65°C$ to $150°C$

NOTES: 1. Voltage values, except interemitter voltage, are with respect to network ground terminal.
2. This is the voltage between two emitters of a multiple-emitter transistor. This rating applies for the '83A only between the following pairs: A1 and B1, A2 and B2, A3 and B3, A4 and B4.

**TTL
MSI**

**TYPES SN5490A, SN5492A, SN5493A, SN54L90, SN54L93,
SN54LS90, SN54LS92, SN54LS93, SN7490A, SN7492A, SN7493A,
'SN74L90, SN74L93, SN74LS90, SN74LS92, SN74LS93
DECADE, DIVIDE-BY-TWELVE, AND BINARY COUNTERS**
BULLETIN NO. DL-S 7611807, MARCH 1974—REVISED OCTOBER 1976

'90A, 'L90, 'LS90 . . . DECADE COUNTERS

**'92A, 'LS92 . . . DIVIDE-BY-TWELVE
COUNTERS**

**'93A, 'L93, 'LS93 . . . 4-BIT BINARY
COUNTERS**

TYPES	TYPICAL POWER DISSIPATION
'90A	145 mW
'L90	20 mW
'LS90	45 mW
'92A, '93A	130 mW
'LS92, 'LS93	45 mW
'L93	16 mW

description

Each of these monolithic counters contains four master-slave flip-flops and additional gating to provide a divide-by-two counter and a three-stage binary counter for which the count cycle length is divide-by-five for the '90A, 'L90, and 'LS90, divide-by-six for the '92A and 'LS92, and divide-by-eight for the '93A, 'L93, and 'LS93.

All of these counters have a gated zero reset and the '90A, 'L90, and 'LS90 also have gated set-to-nine inputs for use in BCD nine's complement applications.

To use their maximum count length (decade, divide-by-twelve, or four-bit binary) of these counters, the B input is connected to the Q_A output. The input count pulses are applied to input A and the outputs are as described in the appropriate function table. A symmetrical divide-by-ten count can be obtained from the '90A, 'L90, or 'LS90 counters by connecting the Q_D output to the A input and applying the input count to the B input which gives a divide-by-ten square wave at output Q_A.

SN54', SN54LS' . . . J OR W PACKAGE
SN54L' . . . J OR T PACKAGE
SN54', SN74L', SN74LS' . . . J OR N PACKAGE

'90A, 'L90, 'LS90 (TOP VIEW)

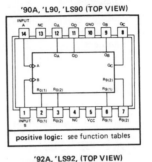

positive logic: see function tables

'92A, 'LS92, (TOP VIEW)

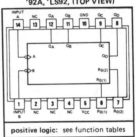

positive logic: see function tables

'93A, 'LS93 (TOP VIEW)

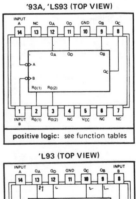

positive logic: see function tables

'L93 (TOP VIEW)

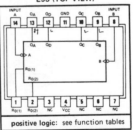

positive logic: see function tables

NC—No internal connection

TYPES SN5490A, '92A, '93A, SN54L90, 'L93, SN54LS90, 'LS92, 'LS93, SN7490A, '92A, '93A, SN74L90, 'L93, SN74LS90, 'LS92, 'LS93 DECADE, DIVIDE-BY-TWELVE, AND BINARY COUNTERS

'90A, 'L90, 'LS90
BCD COUNT SEQUENCE
(See Note A)

COUNT	OUTPUT			
	Q_D	Q_C	Q_B	Q_A
0	L	L	L	L
1	L	L	L	H
2	L	L	H	L
3	L	L	H	H
4	L	H	L	L
5	L	H	L	H
6	L	H	H	L
7	L	H	H	H
8	H	L	L	L
9	H	L	L	H

'90A, 'L90, 'LS90
BI-QUINARY (5-2)
(See Note B)

COUNT	OUTPUT			
	Q_A	Q_D	Q_C	Q_B
0	L	L	L	L
1	L	L	L	H
2	L	L	H	L
3	L	L	H	H
4	L	H	L	L
5	H	L	L	L
6	H	L	L	H
7	H	L	H	L
8	H	L	H	H
9	H	H	L	L

'92A, 'LS92
COUNT SEQUENCE
(See Note C)

COUNT	OUTPUT			
	Q_D	Q_C	Q_B	Q_A
0	L	L	L	L
1	L	L	L	H
2	L	L	H	L
3	L	L	H	H
4	L	H	L	L
5	L	H	L	H
6	H	L	L	L
7	H	L	L	H
8	H	L	H	L
9	H	L	H	H
10	H	H	L	L
11	H	H	L	H

'93A, 'L93, 'LS93
COUNT SEQUENCE
(See Note C)

COUNT	OUTPUT			
	Q_D	Q_C	Q_B	Q_A
0	L	L	L	L
1	L	L	L	H
2	L	L	H	L
3	L	L	H	H
4	L	H	L	L
5	L	H	L	H
6	L	H	H	L
7	L	H	H	H
8	H	L	L	L
9	H	L	L	H
10	H	L	H	L
11	H	L	H	H
12	H	H	L	L
13	H	H	L	H
14	H	H	H	L
15	H	H	H	H

'90A, 'L90, 'LS90
RESET/COUNT FUNCTION TABLE

RESET INPUTS				OUTPUT			
$R_{0(1)}$	$R_{0(2)}$	$R_{9(1)}$	$R_{9(2)}$	Q_D	Q_C	Q_B	Q_A
H	H	L	X	L	L	L	L
H	H	X	L	L	L	L	L
X	X	H	H	H	L	L	H
X	L	X	L	COUNT			
L	X	L	X	COUNT			
L	X	X	L	COUNT			
X	L	L	X	COUNT			

'92A, 'LS92, '93A, 'L93, 'LS93
RESET/COUNT FUNCTION TABLE

RESET INPUTS		OUTPUT			
$R_{0(1)}$	$R_{0(2)}$	Q_D	Q_C	Q_B	Q_A
H	H	L	L	L	L
L	X	COUNT			
X	L	COUNT			

NOTES: A.   Output Q_A is connected to input B for BCD count.
         B.   Output Q_D is connected to input A for bi-quinary count.
         C.   Output Q_A is connected to input B.
         D.   H = high level, L = low level, X = irrelevant

functional block diagrams

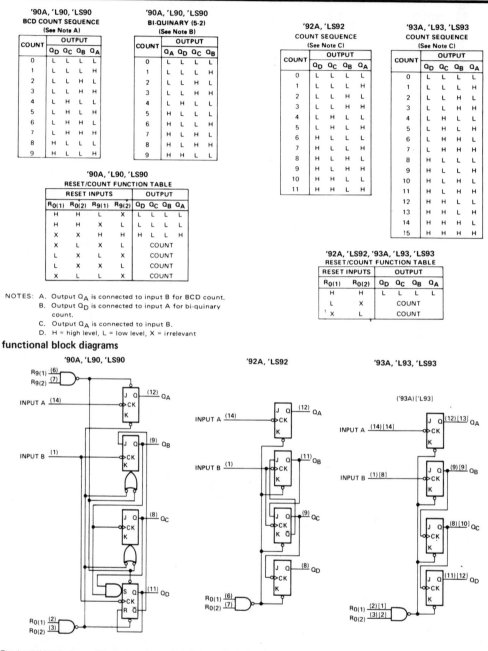

The J and K inputs shown without connection are for reference only and are functionally at a high level.

TTL
MSI

TYPES SN5495A, SN54L95, SN54LS95B, SN7495A, SN74L95, SN74LS95B
4-BIT PARALLEL-ACCESS SHIFT REGISTERS
BULLETIN NO. DL-S 7611872, MARCH 1974—REVISED OCTOBER 1976

TYPE	TYPICAL MAXIMUM CLOCK FREQUENCY	TYPICAL POWER DISSIPATION
'95A	36 MHz	195 mW
'L95	5 MHz	19 mW
'LS95B	36 MHz	65 mW

description

These 4-bit registers feature parallel and serial inputs, parallel outputs, mode control, and two clock inputs. The registers have three modes of operation:

Parallel (broadside) load
Shift right (the direction Q_A toward Q_D)
Shift left (the direction Q_D toward Q_A)

Parallel loading is accomplished by applying the four bits of data and taking the mode control input high. The data is loaded into the associated flip-flops and appears at the outputs after the high-to-low transition of the clock-2 input. During loading, the entry of serial data is inhibited.

Shift right is accomplished on the high-to-low transition of clock 1 when the mode control is low; shift left is accomplished on the high-to-low transition of clock 2 when the mode control is high by connecting the output of each flip-flop to the parallel input of the previous flip-flop (Q_D to input C, etc.) and serial data is entered at input D. The clock input may be applied commonly to clock 1 and clock 2 if both modes can be clocked from the same source. Changes at the mode control input should normally be made while both clock inputs are low; however, conditions described in the last three lines of the function table will also ensure that register contents are protected.

SN5495A, SN54LS95B . . . J OR W PACKAGE
SN7495A, SN74LS95B . . . J OR N PACKAGE
(TOP VIEW)

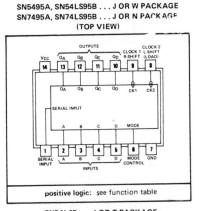

positive logic: see function table

SN54L95 . . . J OR T PACKAGE
SN74L95 . . . J OR N PACKAGE
(TOP VIEW)

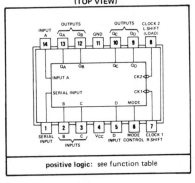

positive logic: see function table

FUNCTION TABLE

INPUTS								OUTPUTS			
MODE	CLOCKS		SERIAL	PARALLEL				Q_A	Q_B	Q_C	Q_D
CONTROL	2 (L)	1 (R)		A	B	C	D				
H	H	X	X	X	X	X	X	Q_{A0}	Q_{B0}	Q_{C0}	Q_{D0}
H	↓	X	X	a	b	c	d	a	b	c	d
H	↓	X	X	Q_B†	Q_C†	Q_D†	d	Q_{Bn}	Q_{Cn}	Q_{Dn}	d
L	L	H	X	X	X	X	X	Q_{A0}	Q_{B0}	Q_{C0}	Q_{D0}
L	X	↓	H	X	X	X	X	H	Q_{An}	Q_{Bn}	Q_{Cn}
L	X	↓	L	X	X	X	X	L	Q_{An}	Q_{Bn}	Q_{Cn}
↑	L	L	X	X	X	X	X	Q_{A0}	Q_{B0}	Q_{C0}	Q_{D0}
↓	L	L	X	X	X	X	X	Q_{A0}	Q_{B0}	Q_{C0}	Q_{D0}
↓	L	H	X	X	X	X	X	Q_{A0}	Q_{B0}	Q_{C0}	Q_{D0}
↑	H	L	X	X	X	X	X	Q_{A0}	Q_{B0}	Q_{C0}	Q_{D0}
↑	H	H	X	X	X	X	X	Q_{A0}	Q_{B0}	Q_{C0}	Q_{D0}

†Shifting left requires external connection of Q_B to A, Q_C to B, and Q_D to C. Serial data is entered at input D.

H = high level (steady state), L = low level (steady state), X = irrelevant (any input, including transitions)
↓ = transition from high to low level, ↑ = transition from low to high level
a, b, c, d = the level of steady-state input at inputs A, B, C, or D, respectively.
Q_{A0}, Q_{B0}, Q_{C0}, Q_{D0} = the level of Q_A, Q_B, Q_C, or Q_D, respectively, before the indicated steady-state input conditions were established.
Q_{An}, Q_{Bn}, Q_{Cn}, Q_{Dn} = the level of Q_A, Q_B, Q_C, or Q_D, respectively, before the most-recent ↓ transition of the clock.

TTL
MSI

TYPES SN54LS138, SN54LS139, SN54S138, SN54S139, SN74LS138, SN74LS139, SN74S138, SN74S139 DECODERS/DEMULTIPLEXERS

BULLETIN NO. DL-S 7611804, DECEMBER 1972–REVISED OCTOBER 1976

- Designed Specifically for High-Speed:
  Memory Decoders
  Data Transmission Systems

- 'S138 and 'LS138 3-to-8-Line Decoders Incorporate 3 Enable Inputs to Simplify Cascading and/or Data Reception

- 'S139 and 'LS139 Contain Two Fully Independent 2-to-4-Line Decoders/ Demultiplexers

- Schottky Clamped for High Performance

TYPE	TYPICAL PROPAGATION DELAY (3 LEVELS OF LOGIC)	TYPICAL POWER DISSIPATION
'LS138	22 ns	32 mW
'S138	8 ns	245 mW
'LS139	22 ns	34 mW
'S139	7.5 ns	300 mW

SN54LS138, SN54S138 . . . J OR W PACKAGE
SN74LS138, SN74S138 . . . J OR N PACKAGE
(TOP VIEW)

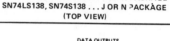
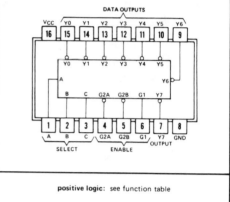

positive logic: see function table

description

These Schottky-clamped TTL MSI circuits are designed to be used in high-performance memory-decoding or data-routing applications requiring very short propagation delay times. In high-performance memory systems these decoders can be used to minimize the effects of system decoding. When employed with high-speed memories utilizing a fast-enable circuit the delay times of these decoders and the enable time of the memory are usually less than the typical access time of the memory. This means that the effective system delay introduced by the Schottky-clamped system decoder is negligible.

The 'LS138 and 'S138 decode one-of-eight lines dependent on the conditions at the three binary select inputs and the three enable inputs. Two active-low and one active-high enable inputs reduce the need for external gates or inverters when expanding. A 24-line decoder can be implemented without external inverters and a 32-line decoder requires only one inverter. An enable input can be used as a data input for demultiplexing applications.

SN54LS139, SN54S139 . . . J OR W PACKAGE
SN74LS139, SN74S139 . . . J OR N PACKAGE
(TOP VIEW)

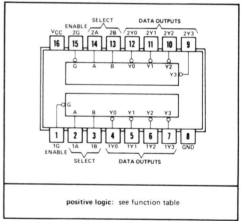

positive logic: see function table

The 'LS139 and 'S139 comprise two individual two-line-to-four-line decoders in a single package. The active-low enable input can be used as a data line in demultiplexing applications.

All of these decoders/demultiplexers feature fully buffered inputs each of which represents only one normalized Series 54LS/74LS load ('LS138, 'LS139) or one normalized Series 54S/74S load ('S138, 'S139) to its driving circuit. All inputs are clamped with high-performance Schottky diodes to suppress line-ringing and simplify system design. Series 54LS and 54S devices are characterized for operation over the full military temperature range of −55°C to 125°C; Series 74LS and 74S devices are characterized for 0°C to 70°C industrial systems.

TYPES SN54LS138, SN54S138, SN54LS139, SN54S139
SN74LS138, SN74S138, SN74LS139, SN74S139
DECODERS/DEMULTIPLEXERS

functional block diagrams and logic

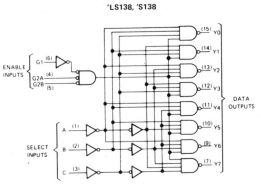

'LS138, 'S138

**'LS138, 'S138
FUNCTION TABLE**

INPUTS				OUTPUTS							
ENABLE		SELECT									
G1	G2*	C	B	A	Y0	Y1	Y2	Y3	Y4	Y5	Y6	Y7
X	H	X	X	X	H	H	H	H	H	H	H	H
L	X	X	X	X	H	H	H	H	H	H	H	H
H	L	L	L	L	L	H	H	H	H	H	H	H
H	L	L	L	H	H	L	H	H	H	H	H	H
H	L	L	H	L	H	H	L	H	H	H	H	H
H	L	L	H	H	H	H	H	L	H	H	H	H
H	L	H	L	L	H	H	H	H	L	H	H	H
H	L	H	L	H	H	H	H	H	H	L	H	H
H	L	H	H	L	H	H	H	H	H	H	L	H
H	L	H	H	H	H	H	H	H	H	H	H	L

*G2 = G2A + G2B

H = high level, L = low level, X = irrelevant

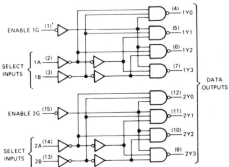

'LS139, 'S139

**'LS139, 'S139
(EACH DECODER/DEMULTIPLEXER)
FUNCTION TABLE**

INPUTS			OUTPUTS			
ENABLE	SELECT					
G	B	A	Y0	Y1	Y2	Y3
H	X	X	H	H	H	H
L	L	L	L	H	H	H
L	L	H	H	L	H	H
L	H	L	H	H	L	H
L	H	H	H	H	H	L

H = high level, L = low level, X = irrelevant

schematics of inputs and outputs

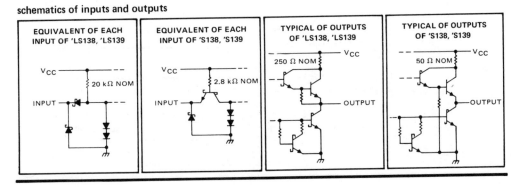

| EQUIVALENT OF EACH INPUT OF 'LS138, 'LS139 | EQUIVALENT OF EACH INPUT OF 'S138, 'S139 | TYPICAL OF OUTPUTS OF 'LS138, 'LS139 | TYPICAL OF OUTPUTS OF 'S138, 'S139 |

TYPES SN54150, SN54151A, SN54152A, SN54LS151, SN54LS152, SN54S151, SN74150, SN74151A, SN74LS151, SN74S151
DATA SELECTORS/MULTIPLEXERS

BULLETIN NO. DL-S 7611819, DECEMBER 1972–REVISED OCTOBER 1976

- '150 Selects One-of-Sixteen Data Sources
- Others Select One-of-Eight Data Sources
- Performs Parallel-to-Serial Conversion
- Permits Multiplexing from N Lines to One Line
- Also For Use as Boolean Function Generator
- Input-Clamping Diodes Simplify System Design
- Fully Compatible with Most TTL and DTL Circuits

TYPE	TYPICAL AVERAGE PROPAGATION DELAY TIME DATA INPUT TO W OUTPUT	TYPICAL POWER DISSIPATION
'150	11 ns	200 mW
'151A	8 ns	145 mW
'152A	8 ns	130 mW
'LS151	11 ns[†]	30 mW
'LS152	11 ns[†]	28 mW
'S151	4.5 ns	225 mW

[†] Tentative data

description

These monolithic data selectors/multiplexers contain full on-chip binary decoding to select the desired data source. The '150 selects one-of-sixteen data sources; the '151A, '152A, 'LS151, 'LS152, and 'S151 select one-of-eight data sources. The '150, '151A, 'LS151, and 'S151 have a strobe input which must be at a low logic level to enable these devices. A high level at the strobe forces the W output high, and the Y output (as applicable) low.

The '151A, 'LS151, and 'S151 feature complementary W and Y outputs whereas the '150, '152A, and 'LS152 have an inverted (W) output only.

The '151A and '152A incorporate address buffers which have symmetrical propagation delay times through the complementary paths. This reduces the possibility of transients occurring at the output(s) due to changes made at the select inputs, even when the '151A outputs are enabled (i.e., strobe low).

SN54150 . . . J OR W PACKAGE
SN74150 . . . J OR N PACKAGE
(TOP VIEW)

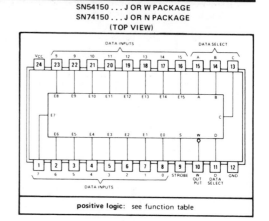

positive logic: see function table

SN54151A, SN54LS151, SN54S151 . . . J OR W PACKAGE
SN74151A SN74LS151, SN74S151 . . . J OR N PACKAGE
(TOP VIEW)

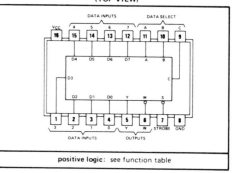

positive logic: see function table

SN54152A, SN54LS152 . . . W PACKAGE
(TOP VIEW)

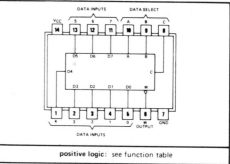

positive logic: see function table

TYPES SN54150, SN54151A, SN54152A, SN54LS151, SN54LS152, SN54S151, SN74150, SN74151A SN74LS151, SN74S151
DATA SELECTORS/MULTIPLEXERS

REVISED OCTOBER 1976

logic

'150

FUNCTION TABLE

INPUTS					OUTPUT
SELECT				STROBE	W
D	C	B	A	S	
X	X	X	X	H	H
L	L	L	L	L	$\overline{E0}$
L	L	L	H	L	$\overline{E1}$
L	L	H	L	L	$\overline{E2}$
L	L	H	H	L	$\overline{E3}$
L	H	L	L	L	$\overline{E4}$
L	H	L	H	L	$\overline{E5}$
L	H	H	L	L	$\overline{E6}$
L	H	H	H	L	$\overline{E7}$
H	L	L	L	L	$\overline{E8}$
H	L	L	H	L	$\overline{E9}$
H	L	H	L	L	$\overline{E10}$
H	L	H	H	L	$\overline{E11}$
H	H	L	L	L	$\overline{E12}$
H	H	L	H	L	$\overline{E13}$
H	H	H	L	L	$\overline{E14}$
H	H	H	H	L	$\overline{E15}$

'151A, 'LS151, 'S151

FUNCTION TABLE

INPUTS				OUTPUTS	
SELECT			STROBE	Y	W
C	B	A	S		
X	X	X	H	L	H
L	L	L	L	D0	$\overline{D0}$
L	L	H	L	D1	$\overline{D1}$
L	H	L	L	D2	$\overline{D2}$
L	H	H	L	D3	$\overline{D3}$
H	L	L	L	D4	$\overline{D4}$
H	L	H	L	D5	$\overline{D5}$
H	H	L	L	D6	$\overline{D6}$
H	H	H	L	D7	$\overline{D7}$

'152A, 'LS152

FUNCTION TABLE

SELECT INPUTS			OUTPUT
C	B	A	W
L	L	L	$\overline{D0}$
L	L	H	$\overline{D1}$
L	H	L	$\overline{D2}$
L	H	H	$\overline{D3}$
H	L	L	$\overline{D4}$
H	L	H	$\overline{D5}$
H	H	L	$\overline{D6}$
H	H	H	$\overline{D7}$

H = high level, L = low level, X = irrelevant
E0, E1 . . . E15 = the complement of the level of the respective E input
D0, D1 . . . D7 = the level of the D respective input

functional block diagrams

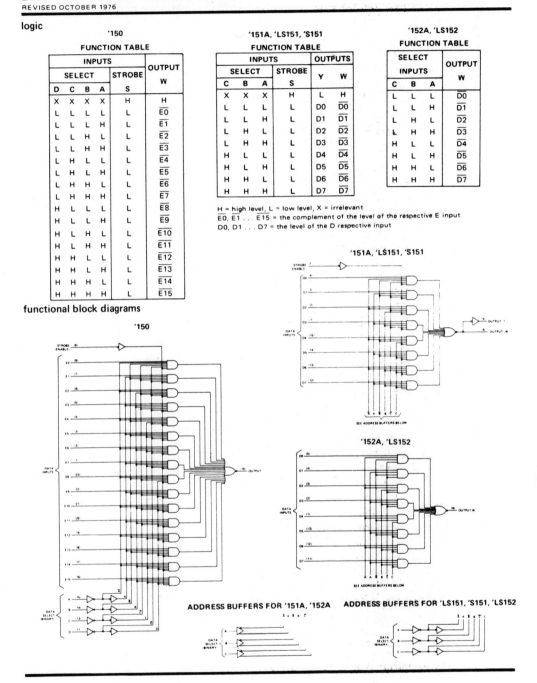

**TTL
MSI**

TYPES SN54153, SN54L153, SN54LS153, SN54S153, SN74153, SN74L153, SN74LS153, SN74S153
DUAL 4-LINE-TO-1-LINE DATA SELECTORS/MULTIPLEXERS
BULLETIN NO. DL-S 7611852, DECEMBER 1972 — REVISED OCTOBER 1976

SN54153, SN54LS153, SN54S153 . . . J OR W PACKAGE
SN54L153 . . . J PACKAGE
SN74153, SN74L153, SN74LS153, SN74S153 . . . J OR N PACKAGE
(TOP VIEW)

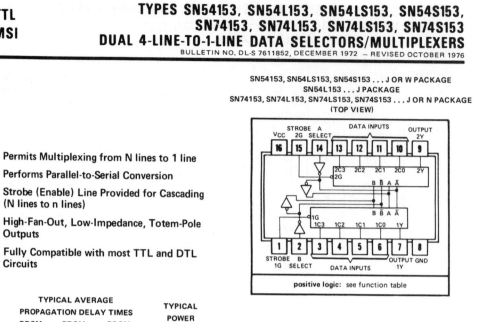

positive logic: see function table

- Permits Multiplexing from N lines to 1 line
- Performs Parallel-to-Serial Conversion
- Strobe (Enable) Line Provided for Cascading (N lines to n lines)
- High-Fan-Out, Low-Impedance, Totem-Pole Outputs
- Fully Compatible with most TTL and DTL Circuits

TYPE	TYPICAL AVERAGE PROPAGATION DELAY TIMES			TYPICAL POWER DISSIPATION
	FROM DATA	FROM STROBE	FROM SELECT	
'153	14 ns	17 ns	22 ns	180 mW
'L153	27 ns	34 ns	44 ns	90 mW
'LS153	14 ns	19 ns	22 ns	31 mW
'S153	6 ns	9.5 ns	12 ns	225 mW

description

Each of these monolithic, data selectors/multiplexers contains inverters and drivers to supply fully complementary, on-chip, binary decoding data selection to the AND-OR-invert gates. Separate strobe inputs are provided for each of the two four-line sections.

FUNCTION TABLE

SELECT INPUTS		DATA INPUTS				STROBE	OUTPUT
B	A	C0	C1	C2	C3	G	Y
X	X	X	X	X	X	H	L
L	L	L	X	X	X	L	L
L	L	H	X	X	X	L	H
L	H	X	L	X	X	L	L
L	H	X	H	X	X	L	H
H	L	X	X	L	X	L	L
H	L	X	X	H	X	L	H
H	H	X	X	X	L	L	L
H	H	X	X	X	H	L	H

Select inputs A and B are common to both sections.
H = high level, L = low level, X = irrelevant

absolute maximum ratings over operating free-air temperature range (unless otherwise noted)

Supply voltage, V_{CC} (see Note 1) . . . . 7 V
Input voltage: '153, 'L153, 'S153 . . . . 5.5 V
        'LS153 . . . . 7 V
Operating free-air temperature range: SN54', SN54L', SN54LS, SN54S' Circuits . . . . −55°C to 125°C
        SN74', SN74L', SN74LS', SN74S' Circuits . . . . 0°C to 70°C
Storage temperature range . . . . −65°C to 150°C

NOTE 1: Voltage values are with respect to network ground terminal.

TYPES SN54157, SN54L157, SN54LS157, SN54LS158, SN54S157, SN54S158, SN74157, SN74L157, SN74LS157, SN74LS158, SN74S157, SN74S158 QUADRUPLE 2-LINE-TO-1-LINE DATA SELECTORS/MULTIPLEXERS
BULLETIN NO. DL-S 7711847, MARCH 1974—REVISED AUGUST 1977

features

- Buffered Inputs and Outputs
- Three Speed/Power Ranges Available

TYPES	TYPICAL AVERAGE PROPAGATION TIME	TYPICAL POWER DISSIPATION
'157	9 ns	150 mW
'L157	18 ns	75 mW
'LS157	9 ns	49 mW
'S157	5 ns	250 mW
'LS158	7 ns	24 mW
'S158	4 ns	195 mW

applications

- Expand Any Data Input Point
- Multiplex Dual Data Buses
- Generate Four Functions of Two Variables (One Variable Is Common)
- Source Programmable Counters

description

These monolithic data selectors/multiplexers contain inverters and drivers to supply full on-chip data selection to the four output gates. A separate strobe input is provided. A 4-bit word is selected from one of two sources and is routed to the four outputs. The '157, 'L157, 'LS157, and 'S157 present true data whereas the 'LS158 and 'S158 present inverted data to minimize propagation delay time.

SN54157, SN54LS157, SN54S157 . . . J OR W PACKAGE
SN54L157 . . . J PACKAGE
SN74157, SN74L157, SN74LS157, SN74S157 . . . J OR N PACKAGE
(TOP VIEW)

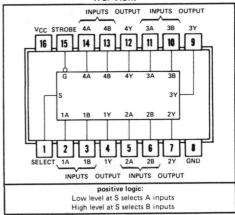

positive logic:
Low level at S selects A inputs
High level at S selects B inputs

SN54LS158, SN54S158 . . . J OR W PACKAGE
SN74LS158, SN74S158 . . . J OR N PACKAGE
(TOP VIEW)

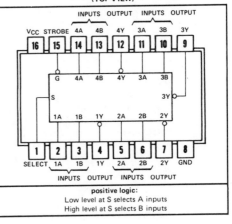

positive logic:
Low level at S selects A inputs
High level at S selects B inputs

FUNCTION TABLE

INPUTS				OUTPUT Y	
STROBE	SELECT	A	B	'157, 'L157, 'LS157, 'S157	'LS158 'S158
H	X	X	X	L	H
L	L	L	X	L	H
L	L	H	X	H	L
L	H	X	L	L	H
L	H	X	H	H	L

H = high level, L = low level, X = irrelevant

absolute maximum ratings over operating free-air temperature range (unless otherwise noted)

Supply voltage, V_{CC} (see Note 1)	7 V
Input voltage: '157, 'L157, 'S158	5.5 V
'LS157, 'LS158	7 V
Operating free-air temperature range: SN54', SN54L', SN54LS', SN54S' Circuits	−55°C to 125°C
SN74', SN74L', SN74LS', SN74S' Circuits	0°C to 70°C
Storage temperature range	−65°C to 150°C

NOTE 1: Voltage values are with respect to network ground terminal.

SYNCHRONOUS 4-BIT COUNTERS

160   DECADE, DIRECT CLEAR

161   BINARY, DIRECT CLEAR

162   DECADE, SYNCHRONOUS CLEAR

163   BINARY, SYNCHRONOUS CLEAR

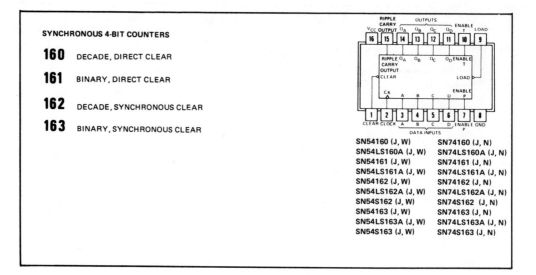

SN54160 (J, W)   SN74160 (J, N)
SN54LS160A (J, W)   SN74LS160A (J, N)
SN54161 (J, W)   SN74161 (J, N)
SN54LS161A (J, W)   SN74LS161A (J, N)
SN54162 (J, W)   SN74162 (J, N)
SN54LS162A (J, W)   SN74LS162A (J, N)
SN54S162 (J, W)   SN74S162 (J, N)
SN54163 (J, W)   SN74163 (J, N)
SN54LS163A (J, W)   SN74LS163A (J, N)
SN54S163 (J, W)   SN74S163 (J, N)

appendix C

References

Blakeslee, Thomas R. *Digital Design with Standard MSI and LSI.* 2d ed. New York: Wiley, 1979.

Greenfield, Joseph D. *Practical Digital Design Using IC's.* New York: Wiley, 1977.

Heiserman, David L. *Handbook of Digital IC Applications.* Englewood Cliffs, N.J.: Prentice-Hall, 1980.

Hill, Frederick J. and Peterson, Gerald R. *Digital Systems Hardware Organization and Design.* 2d ed. New York: Wiley, 1978.

Lancaster, Don. *TTL Cookbook.* Indianapolis: Howard W. Sams, 1981.

Lewin, Morton H. *Logic Design and Computer Organization.* Reading, Mass.: Addison-Wesley, 1983.

Mano, M. Morris. *Computer Logic Design.* Englewood Cliffs, N.J.: Prentice-Hall, 1972.

Mowle, Frederick J. *A Systematic Approach to Digital Logic Design.* Reading, Mass.: Addison-Wesley, 1977.

Newell, Sydney B. *Introduction to Microcomputing.* New York: Harper & Row, 1982.

Peatman, John B. *Digital Hardware Design.* New York: McGraw-Hill, 1980.

Sandige, Richard S. *Digital Concepts Using Standard Integrated Circuits.* New York: McGraw-Hill, 1978.

Signetics Corporation. *TTL Logic Data Manual 1982.* Sunnyvale, Calif: Signetics, 1982.

Stone, Harold S. *Microcomputer Interfacing.* Reading, Mass.: Addison-Wesley, 1982.

Taub, Herbert. *Digital Circuits and Microprocessors.* New York: McGraw-Hill, 1982.

Texas Instruments, Inc. *The TTL Data Book for Design Engineers.* 2d ed. Dallas: Texas Instruments, 1976.

Texas Instruments, Inc. *1981 Supplement to the TTL Data Book for Design Engineers.* 2d ed. Dallas: Texas Instruments, 1981.

appendix
D

Recommended Parts List and Lab Equipment

The following parts will support one lab group (ideally two students):

Quantity	Type	Description
4	7400	quad two-input NAND gate
4	7402	quad two-input NOR gate
4	7404	hex inverter
2	7408	quad two-input AND gate
3	7410	triple three-input NAND gate
2	7420	dual four-input NAND gate
2	7432	quad two-input OR gate
2	7448	BCD to seven-segment decoder
2	7483	4-bit binary full-adder
4	7474	dual D-type flip-flops
4	7476	dual J-K flip-flops
2	7493	4-bit binary counters
2	7490	decade counters
4	7495	4-bit shift registers
2	74LS138	3-to-8 line decoder/mux
2	74LS139	dual 2-to-4 line decoder/mux
2	74151	1-of-8 data selector/mux
2	74153	dual 4-line to 1-line data selector/mux
2	74157	quad 2-line to 1-line data selector/mux
1	MAN74	Common cathode seven-segment display

Test Equipment (per five groups)

- 1 logic probe (Global Specialties, LP-2, or equivalent)

Assorted Equipment and Supplies

- 500-foot hookup wire, single strand, 22 gauge, assorted colors
- 2 integrated circuit pullers
- 4 pairs of wire cutters and strippers

Logic Circuit Prototyping Board

A prototyping board similar in capabilities to the Digi-Designer manufactured by E & L Instruments is required for each lab group. The Digi-Designer is a complete digital-circuit design instrument that handles both IC and discrete components without soldering. Connections are made with solid 22 gauge wire. The device has a regulated 5 volt power supply, a pulse generator (clock) providing continuous digital pulses in discrete steps from 1 Hz to 100 kHz, two debounced pushbuttons, four undebounced switches for applying logic levels, four LED lamp monitors, and a standard component mounting socket with room for up to eight 14-pin standard packages.

Index